Contents

	Foreword	xvii
	Author's Preface	xix
1	*Factors which have influenced modern bridge design*	1
	Welded steel construction	1
	Prestressed concrete construction	7
	Mossband Bridge	14
2	*New erection methods*	20
	New foundation construction methods	20
	Moving bridges into position, by rolling and sliding	25
3	*Some notable steel bridges in the UK*	30
	Grosvenor Bridge, S Region	30
	Trowell Bridge, LM Region	35
	The Retford 'Dive-under bridge', E Region	37
	Tinsley Canal bridge, E Region	37
	Deptford Creek Vertical lift bridge, S Region	40
	Three Western Region bridges	43
	Churchdown	43
	Haresfield	46
	Hyde Lane	48
	Fledborough Viaduct, E Region	49
4	*Some notable concrete bridges in the UK*	57
	The Adam Viaduct, LM Region	57
	The River Esk Viaduct, LM Region	58
	Three bridges at Manchester, LM Region	59
	Stockport Road	60
	Hyde Road	62
	Fairfield Street	63
	The Penrith Bridges, LM Region	65
	Besses o' th' Barn, Manchester, LM Region	69
	The Frodsham Bridges 4A, 4B, 5, LM Region	74
	Leeds Central, Canal Bridge, E Region	78
	Wandsworth Bridge No 3A, S Region	78
	The Rugby Fly-over, LM Region	80
	The Bletchley Fly-over, LM Region	80

Pilgrim Street, Newcastle upon Tyne, E Region 81

Stevenage No 85A, E Region 84

Bridge No 1E, Derby, LM Region 90

Pipe Bridge, Runcorn, LM Region 92

Roofing over cuttings 93

5 *Footbridges on British Railways* 97

6 *Subways on British Railways* 115

7 *Some notable bridges overseas* 122

Central Africa: River Nile bridge at Pakwash 122

China: River Yangtze bridge at Nanking 122

India: Reconstructed bridge over the Juggal River 123

Pakistan: The Attock Bridge over the River Indus 124

Burma: The Sittang River Bridge 127

Japan: The River Arakawa Bridge 130

Austria: Two Bridge renewals on the Arlberg line 134

Portugal: The River Tagus suspension bridge at Lisbon 137

France 138

Bridge No C14 at Rungis market 138

Viaduct No C F 6 Rungis market approach 139

The Caronte swing bridge 141

Denmark 143

The bridge over the Kleinenbelt 143

Switzerland 143

The Kander Viaduct, Luogelkin Viaduct, Landwasser Viaduct, Gstaad Viaduct, Bietschtal Bridge, Lorraine Viaduct, Grandfey Bridge 143

Germany 147

Fehmarnsund Bridge 147

Index 156

List of Illustrations

Plates

1.1   Bridge No 125. B.R. Eastern Region   2

1.2   Bridge No 123. B.R. Scottish Region   3

1.3   German 'half-through' steel bridge   8

1.4   Bridge No 66 Watford. B.R. London Midland Region   11

1.5   Bridge No 60A Atherstone. B.R. London Midland Region   12

1.6   Bridge No 37: rebuilding. B.R. London Midland Region   13

1.7   Rebuilt arch bridge. SNCF Region de l'Est   14

1.8   Raised bridge. SNCF Region de l'Est   15

1.9   Formwork for 'table spans' of Mossband Bridge. B.R. London Midland Region   16

1.10   Completed 'table spans' of Mossband Bridge. B.R. London Midland Region   17

1.11   Placing beams for 'suspended spans' of Mossband Bridge. B.R. London Midland Region   18

1.12   Completed Mossband Bridge. B.R. London Midland Region   19

2.1   Stevenage Bridge No 85A: sheet piling at site. B.R. Eastern Region   20

2.2   Stevenage Bridge No 85A: piling on middle tracks. B.R. Eastern Region   21

2.3   Churchdown Bridge: details of sledges. B.R. Western Region   25

2.3a   Churchdown Bridge: rolling-in completed. B.R. Western Region   25

2.4a–c   Sliding-in trials at Crewe. B.R. London Midland Region   26–8

3.1   Grosvenor Bridge rebuilding. B.R. Southern Region   30

3.2   Grosvenor Bridge rebuilding: river works. B.R. Southern Region   32

3.3   Grosvenor Bridge rebuilding: floating away old arch ribs. B.R. Southern Region   33

3.4   Grosvenor Bridge rebuilding: erecting new ribs by service girder.   33

3.5   Trowell Bridge. B.R. London Midland Region   39

3.6   Retford dive-under bridge. B.R. Eastern Region   41

3.7   Tinsley Canal Bridge. B.R. Eastern Region   42

3.8   Deptford Creek vertical lift bridge. B.R. Southern Region   43

3.9   Churchdown Bridge. B.R. Western Region   45

3.10   Churchdown Bridge: excavation for portal frame legs. B.R. Western Region   46

3.11   Churchdown Bridge: 'pipe-jacked' tube for cross member. B.R. Western Region   47

3.12   Haresfield Bridge. B.R. Western Region   47

3.13   Hyde Lane Bridge. B.R. Western Region   55

3.14   Fledborough Viaduct reconstruction. B.R. Eastern Region   56

4.1   Adam Viaduct. B.R. London Midland Region   57

4.2   River Esk Viaduct. B.R. London Midland Region   58

4.3   Fairfield Street Bridge, Manchester. B.R. London Midland Region   62

4.4   Clifton Bridge, Penrith. B.R. London Midland Region   67

4.5   Clifton Bridge under construction. B.R. London Midland Region   70

4.6   Clifton Bridge after sliding-in operation. B.R. London Midland Region   71

4.7   Besses o' th' Barn Bridge, Manchester. B.R. London Midland Region   71

4.8   Besses o' th' Barn Bridge; end view. B.R. London Midland Region   72

4.9   Besses o' th' Barn Bridge; assembly of main 'spine'. B.R. London Midland Region   74

4.10 Frodsham Bridge No 4A. B.R. London Midland Region   75

4.11 Frodsham Bridge No 4A: end view. B.R. London Midland Region   76

4.12 Canal Bridge No 11; Leeds City Station. B.R. London Midland Region   78

4.13 Rugby Fly-over. B.R. London Midland Region   80

4.14 Bletchley Fly-over. B.R. London Midland Region   81

4.15 Pilgrim Street Bridge, Newcastle upon Tyne. B.R. Eastern Region   83

4.16 Stevenage Bridge No 85A. B.R. Eastern Region   85

4.17 Bridge No 1E at Derby; under construction. B.R. London Midland Region   86

4.18 Bridge No IE at Derby: prior to sliding-in. B.R. London Midland Region   87

4.19 Bridge No 1E at Derby: completed. B.R. London Midland Region   90

4.20 Pipe bridge at Runcorn. B.R. London Midland Region   92

4.21 Roofing over Edge Hill cutting. B.R. London Midland Region   93

4.22 New roof beams for New Street Station, Birmingham. B.R. London Midland Region   94

4.23 Reconstruction of Hampstead Road bridge during Euston Station remodelling.
      B.R. London Midland Region   95

4.24 Bridge No 3A, Wandsworth. B.R. Southern Region   96

5.1   Adamstown Footbridge, Cardiff. B.R. Western Region   97

5.2   Congleton Footbridge. B.R. London Midland Region   98

5.3   Temple Mills Footbridge: tubular 'Usk' type by Tubewrights. B.R. Eastern Region   99

5.4   Pudsey Footbridge: Tubewrights hollow steel section type. B.R. Eastern Region   100

5.5   Scunthorpe Footbridge: prefabricated units. B.R. Eastern Region   101

5.6   Scunthorpe Footbridge: view of deck. B.R. Eastern Region   101

5.7   Exeter Footbridge: concrete unit construction   102

5.8   Precast units in maker's yard   103

5.9   Musgrave Hospital Footbridge, Belfast   104

5.10 Footbridge at York. B.R. Eastern Region   105

5.11 Footbridge No 89: raised for electrification   106

5.12 Footbridge at Prestbury: reconstructed for electrification. B.R. London Midland
      Region   107

5.13 Footbridge at Tring: unit construction. B.R. London Midland Region   108

5.14 Footbridge at Ellesmere Port. B.R. London Midland Region   109

5.15 Footbridge at Abbey Wood. B.R. Southern Region   110

5.16 Footbridge at Marazion. B.R. Western Region   111
5.17 Footbridge at Tenby. B.R. Western Region   112
5.18 Footbridge at Bishop's Stortford. B.R. Eastern Region   113
5.19 Footbridge at Westcliff. B.R. Southern Region   114
6.1   'Pipe jacking': looking down into a drive pit   116
6.2   'Pipe jacking': twin pipe sheilds, Exeter   117
6.3   Pedestrian subway at Ash. B.R. Southern Region   118
6.4   Watford vehicle subway: jack assembly. B.R. London Midland Region   119
6.5   Watford vehicle subway: driving operations. B.R. London Midland Region   119
7.1   River Nile Bridge at Pakwash, Uganda   122
7.2   River Yangtze Bridge, China: General view   125
7.3   River Yangtze Bridge, China: opening ceremony   125
7.4   River Juggal Bridge, India: bank collapse   126
7.5   River Juggal Bridge, India: placing the new girder   126
7.6   Attock Bridge, River Indus, Pakistan   127
7.7   Sittang River Bridge, Burma   130
7.8   Arakawa River Bridge, Japan   131
7.9   Arakawa River Bridge, Japan: end view   134
7.10 Schnatobel Bridge renewal, Austria   135
7.11 Otztaler Bridge renewal, Austria   135
7.12 River Tagus Bridge, Portugal   136
7.13 Bridge No C14, France   138
7.14 Bridge No CF6, France   141
7.15 Caronte swing bridge reconstruction, France   141
7.16 Caronte swing bridge completed, France   142
7.17 Kleinenbelt Bridge, Denmark   142
7.18 Kander Viaduct, Switzerland   146
7.19 Luogelkin Viaduct, Switzerland   147
7.20 Landwasser Viaduct, Switzerland   148
7.21 Gstaad Viaduct, Switzerland   148
7.22 Bietschtel Bridge, Switzerland   149
7.23 Lorraine Viaduct, Berne, Switzerland   150
7.24 Grandfey Viaduct, Switzerland   151
7.25 Fehmarnsund bridge, Germany/Denmark   152

Drawings and Plans

1 Bridge No 125. Concrete well deck on steel girders. B.R. Eastern Region   4–5
2 Bridge No 123. Welded steel 'half through' type. B.R. Scottish Region   6

3 Standard B.R. 'half through' type Z bridge 8–9

4 German 'half-through' welded steel bridge 9

5 Buck Lane Bridge: prestressed concrete. B.R. Eastern Region 10

6 Ipswich Road Bridge, Colchester: prestressed concrete. 10

7 Mouselow Bridge: prestressed concrete. B.R. Eastern Region 10

8 Mossband Bridge; prestressed concrete overbridge. B.R. London Midland Region 16

9 Bridge No 85A Stevenage: sheet piling for foundations. B.R. Easter Region 22–3

10 Alphon Brook Bridge: 'pipe-jacked' foundations. B.R. Western Region 24

11 Sliding-in sledge for Clifton bridge. B.R. London Midland 29

12 Grosvenor Bridge. Old and new bridges. B.R. Southern Region 31

13 Grosvenor Bridge reconstruction: foundation pier plan. B.R. Southern Region 31

14 Grosvenor Bridge reconstruction: elevation of new pier. B.R. Southern Region 34

15 Grosvenor Bridge reconstruction: arch rib elevation. B.R. Southern Region 34

16 Grosvenor Bridge reconstruction, arch rib cross section. B.R. Southern Region 35

17 Trowell Bridge: cross section. B.R. London Midland Region 36

18 Retford dive-under bridge: site plan. B.R. Eastern Region 37

19 Retford dive-under bridge: plan and cross-section. B.R. Eastern Region 38

20 Tinsley canal bridge: details. B.R. Eastern Region 40

21 Deptford Creek lifting bridge: elevation and plan. B.R. Southern Region 44

22 Churchdown Bridge: outline diagram. B.R. Western Region 46

23 Haresfield Bridge: outline diagram. B.R. Western Region 48

24 Fledborough Viaduct: old bridge. B.R. Eastern Region 50–1

25 Fledborough Viaduct: reconstruction. B.R. Eastern Region 52–3

26 Fledborough Viaduct: new bridge, cross sections. B.R. Eastern Region 54

27 Stockport Road Bridge: elevation and section. B.R. London Midland Region 59

28 Hyde Road Bridge: elevation and section. B.R. London Midland Region 60–1

29 Fairfield Street Bridge: section. B.R. London Midland Region 63

30 Fairfield Street Bridge: plan. B.R. London Midland Region 64–5

31 Fairfield Street Bridge: elevation and column 'D'. B.R. London Midland Region 66

32 Clifton Bridge: elevation and section. B.R. London Midland Region 68

33 Penrith Station Bridge: section. B.R. London Midland Region 68

34 Skirsgill Bridge: section. B.R. London Midland Region 69

35 Besses o' th' Barn Bridge: elevation. B.R. London Midland Region 70

36 Besses o' th' Barn Bridge: sections. B.R. London Midland Region 73

37 Frodsham Bridges No 4A and 5: plan and elevation. B.R. London Midland Region 74

38 Frodsham Bridges No 4A and 5: method of erection. B.R. London Midland Region 77

39 Frodsham Bridges No 4B: section. B.R. London Midland Region 77

40 Bridge No 3A at Wandsworth: isometric drawing. B.R. Southern Region 79

41 Pilgrim Street Bridge: method of construction. B.R. Eastern Region 82

42 Bridge No 85A at Stevenage: details. B.R. Eastern Region 84

43 Bridge No 1E at Derby: diagram and sections. B.R. Eastern Region 86–7

44 Bridge No 1E at Derby: detail of foundation. B.R. Eastern Region 88–9

45 'Seerthrust' pipe-jacking system 115

46 Richmond pedestrian subway 117

47 Watford car subway: plan and longitudinal section. B.R. London Midland Region 120–1

48 Watford car sub-way: cross section. B.R. London Midland Region 120

49 Bridge over the Yangtsi River: site plan 123

50 River Juggal Bridge: site plan 124

51 River Juggal Bridge: method of reconstruction 124

52 Sittang Bridge: site plan 127

53 Attock Bridge: elevation and plan 128–9

54 Arakawa Bridge: elevation and plan 131

55 Arakawa Bridge: section 132–3

56 River Tagus bridge: elevation 136–7

57 Paris Rungis Market Bridge No C14: elevation and plan 139

58 Paris Rungis Market Bridge No C14: cross section 139

59 Paris Rungis Market Bridge No C14: details 140

60 Kleinenbelt Bridge and Viaduct, Denmark: detail of approach span 143

61 Kleinenbelt Bridge and Viaduct, Denmark 144–5

62 Fehmarnsund Bridge: detail of approach span 152

63 Fehmarnsund Bridge: detail of piers 153

64 Fehmarnsund Bridge: detail of main span 154–5

Acknowledgments

The author's thanks are due to the following for the assistance with photographs, drawings and plans:

W. S. Atkins and Partners: Plates 3.8, 4.15. Figure 41

British Railways: Plates 1.1, 1.2, 1.4, 1.5, 1.6, 1.9, 1.10, 2.1, 2.2, 2.3, 2.4, 2.5, 2.6, 3.5, 3.6, 3.7, 3.9, 3.10, 3.11, 3.12, 3.13, 4.1 to 4.14, 4.16 to 4.20, 5.1, 5.2, 5.7, 5.10 to 5.19, 6.4, 6.5. Figures 1 to 3, 9, 47

British Steel Corporation: Plates 5.3, 5.4

Concrete: Plates 1.9 to 1.12, 5.8, 5.9, 7.15. Figures, 8, 10, 35, 36, 42 to 44

Eisenbahn Technische Rundschau: Figures 60 to 64

Engineering: Figure 40

Freeman Fox and Partners: Plates 3.1 to 3.4. Figure 12

Institution of Civil Engineers: Plates 7.15, 7.16. Figures 5 to 7, 27 to 31

Japanese National Railways: Plates 7.8, 7.9

Pier Headings Ltd.: Plate 4.21

Railway Gazette: Plates 1.7 to 1.12, 7.1 to 7.5, 7.10, 7.11, 7.13, 7.14. Figures 11, 13 to 23, 32 to 34, 37 to 39, 49 to 51, 53 to 59

William J. Rees: Plates 6.1, 6.2. Figure 45

Rendel, Palmer and Tritton: Plates 3.14, 5.5, 5.6, 7.6, 7.7. Figures 24 to 26, 52

Revue Générale des Chemins de Fer: Plates 7.15, 7.16

Société Nationale des Chemins de Fer Française: Plates 1.7, 1.8, 7.13, 7.14

Swiss Federal Railways: Plates 7.18 to 7.24

Tube Headings: Plate 6.3. Figure 46

I wish to acknowledge with thanks the very considerable help given me from time to time by the Chief Civil Engineer's Departments of the several regions of British Railways and their staffs, in the preparation of the original articles which form the basis of the book, as well as the officers of overseas railways such as France, Japan and Switzerland. The bulk of the drawings in the book were prepared originally by the *Railway Gazette* from material obtained from British and other railways, and most of the photographs also came from the same sources.

I would also like to thank the library of the Institution of Civil Engineers, Messrs Freeman Fox and Partners, and Rendel Palmer and Tritton for permission to make use of photographs and plans, and the latter have also helped with the text.

My especial thanks are also due to the late managing editor of the *Railway Gazette*, Mr H. M. Dannatt, for his encouragement, help and advice.

Foreword

by F. R. L. Barnwell, OBE, ERD, FICE, MIQ
formerly Chief Civil Engineer, Western Region, British Railways

Having known F. A. W. Mann for some thirty years, from the time when he was on the London and North Eastern Railway, it has given me considerable pleasure to read this interesting and instructive book.

One of the railway engineer's chief problems has always been, of course, bridge erection with the least interference to traffic—not always easy—and, therefore, the chapters on pipe jacking, particularly for abutments, are of especial value, not only to the engineer of a railway but all associated with transportation.

The developments in design are ably brought out by the author so that the book makes most informative reading, and by avoiding extensive calculations is of a wider interest to less specialist readers.

Author's Preface

For a long time, I have wanted to write a book about railways. As a civil engineer, I have been connected professionally with railways all my life, first with consulting engineers specialising in railway electrification, followed by ten years with the former London and North Eastern Railway, and finally ending up again with consultants engaged with railway construction and planning. After my retirement I started writing technical articles for the *Railway Gazette*, and one of the first was on the reconstruction of the Grosvenor Bridge out of Victoria Station. These articles were well received and although there are other aspects of railway engineering on which I could have written, I became convinced that railway bridges must be my subject.

There is a certain amount of romance about bridges. Of all the more utilitarian works of mankind, it could be maintained that the bridge is one which has most captured the admiration of the citizen, and the imagination of the artist. Possibly this is in part due to the fact that bridges have, of necessity, to be built in places where there is natural beauty, such as valleys and rivers. They also tend to survive longer than other works, since they must possess robustness of construction and are so much used that they receive adequate maintenance. It could be reasonably argued that it is hard to design an ugly bridge.

While railway bridges do not always provide the most elegant examples of the craft, there are many which, by their massive proportions and obvious strength, strike the imagination. The bridges in the west of England designed by Brunel, Robert Stephenson's Britannia Bridge across the Menai Straits, the Tay Bridge, and above all the Forth Bridge, are famous not only among engineers, but also to the general public. The Forth Bridge is still, perhaps, the most famous railway bridge ever built, and even if, with its massive proportions, and spectacular appearance, it could hardly be described as elegant, it certainly has grandeur.

As the number of my articles on modern railway bridges, which were published not only in the *Railway Gazette*, but in other journals such as *Concrete*, increased, I became more than ever convinced that they could form the basis of a useful book. Quite a lot of rather laborious work is involved in the preparation of such articles, particularly with foreign railways, and one realises how soon after they have been written they tend to be forgotten and lie buried in the back numbers of the journals in which they originally appeared. This, it seems to me, makes a further reason why a book such as this can be useful by the collection under one cover of the records of bridges for future reference. It also became clear to me that very few books exist on railway bridges only.

The recent publication of Mr P. S. A. Berridge's book *The Girder Bridge since Brunel and Others* was also a stimulus, since this fascinating work deals with steel bridges only and is fairly historical. My book deals with both steel and concrete bridges of relatively recent construction, and an international survey has been attempted, although due to my personal experience the main emphasis is on practice in the UK. It might well be that a second volume to include the latest bridges worthy of note and other examples of overseas designs which I have been unable to include here could be written at a future date.

Since 1950, after the end of the second world war, great technological and scientific advances have taken place which have strongly influenced bridge design throughout the world, particularly in the use of steel and concrete.

In those countries such as France, Italy and Germany in Europe, and India and Burma in Asia, where there was considerable destruction of the railway systems, a large number of bridges had to be rebuilt, and advantage was taken of the new engineering techniques in their reconstruction.

At the same time, new thinking on transportation, and the place of railways in relation to other forms of transport, such as road and air services, has resulted in alterations to the railway systems which has often called for new bridges.

In the case of new emergent countries which have now achieved their independence, the natural desire to become 'up to date' has often involved large civil engineering development schemes which embrace new or improved railway facilities and connections for which new modern design bridges have been required.

The conditions under which new railway bridges have been called for naturally differ according to the circumstances existing in each country, and in Britain the factors are generally rather unique. This is another reason why the emphasis of the book tends to be on the United Kingdom.

Some of the designs produced here have already aroused interest in other countries, notably, and recently, in Japan, where a new bridge being constructed at the time of writing at Derby has been the object of interest to the civil engineers of the Japanese National Railways.

The book is not a technical treatise on bridge design, nor is it a mere catalogue. I feel, however, that there is ample technical detail provided to make it of interest to railway civil engineers and yet at the same time remain readable to those interested in railways in a more general way. It also provides in one cover a partial record of recent practice in bridge design, which I hope may be amplified in a second volume.

F. A. W. MANN

I Factors which have influenced modern bridge design

Since the end of the second world war two main factors have made necessary the construction of a number of new bridges on the railways of the United Kingdom, and some remarkable and original designs have been produced. These two factors are, first, large-scale electrification on the overhead-line system, and, second, the continuing construction of the national motorway network. There is also the reshaping of the whole railway system leading to track realignments and rearrangements, and increased loading on existing bridges which has necessitated some fairly large-scale reconstructions of older bridges as well as new bridges.

The approach to design has also been radically changed by the establishment of the two techniques of welded steel and prestressed concrete construction in British railway engineering practice. Very naturally, railway engineers have to be conservative, and are not justified in adopting new techniques on a big scale, before they have been proved beyond all doubt and in fact some early troubles with prestressed concrete on the Southern Region caused a setback in its application. It is not really idle boasting to claim that safety on British Railways is second to no other country, and this is largely due to the sound technical approach of the Civil Engineering Departments.

In general, prestressed concrete bridges have been favoured mostly on the London Midland Region, while on the Eastern and Western Regions welded steel construction is the more general use, but there is no hard-and-fast rule about this.

The construction of the motorways has to some extent affected the north-western side of the country more than the eastern side and this may be partially explained by the fact that the terrain being more hilly, the new roads tend to pass under railways, whereas on the relatively flatter eastern side of the country the roads can more readily be made to cross over, so that few new rail bridges are required. It is, perhaps, for this reason that the more interesting and original bridge designs have emerged from the London Midland Region during the past thirty years. Nevertheless, a number of fine designs are to be found on the other Regions and some examples are included in this book. These new bridges tend to be large on account of the width of the roads, which also often have to cross under the railway on the skew. Indeed, had it not been for the motorways, it is doubtful if some of the bridges described would have ever seen the light of day, and new bridgework would have been confined to renovations and reconstructions for road widening or strengthening to give increased capacity.

Welded steel construction

Among examples of welded construction of greater importance must be placed the reconstruction of Grosvenor Bridge on the Southern Region of British Railways, which was a particularly difficult and complex operation, and the Canal Bridge at Tinsley on the Eastern Region, which was, at the time it was built, the largest single all-welded span in the country.

A more recent welded steel bridge is that at Churchdown on the Western Region, which shows considerable ingenuity in the methods of construction and erection. This bridge was required on account of a motorway, as also were two other steel bridges on the same Region at Haresfield and Hyde Lane, which are described in the chapter on steel bridges, none of which was occasioned by electrification, only a few by motorways, and the rest on account of age, accidents, or the necessity to increase their capacity.

Although welding had been used on British Railways before the war, it did not come to be used on a big scale until afterwards. In 1948 seven bridges on the Scottish Border were washed away completely in a severe flood. Six of them were replaced by bridges of deck-type construction with a reinforced concrete well deck keyed to and supported by welded steel plate girders as shown in fig 1, showing Bridge No 125.

The seventh, No 123, was replaced by a welded steel 'half-through'-type shown in fig 2. It consists of main girders built of welded steel plates to form an I section with a deck formed of transverse steel joists encased in concrete. This bridge may be said to have set the pattern for a series of five standard 'half-through'-type bridges which have been evolved by the Civil Engineers Department of the British Railways Board, and are generally similar in cross-section to the left-hand side of fig 2. They are suitable for spans from 15 to 34 m and can be arranged for straight or skew positions. The smallest type Z is shown in fig 3. This is suitable for spans up to 15 m, and differs from the others in that the main girders

Plate 1.1   *Bridge No. 125, B.R. Eastern Region*

Plate 1.2   *Bridge No. 123, B.R. Scottish Region*

are built up to form a Z section the outward splay of which gives greater side clearances. It is also designed to be lifted into position by crane, while the other types B, C, D, and E, are designed to be rolled into position.

Where rolling-in is not practicable, modified types B1, C1, D1, and E1 have a steel battle deck substituted for the composite steel and concrete deck which is attached to the side girders by high tensile bolts, so that the bridge can be lifted into position in sections. Type E or E1 is the largest, and is suitable for spans up to 34 m.

The system of forming a composite bridge with steel main girders and a reinforced con-crete deck similar to Bridge No 125 in fig 1 has been used on many bridges, and the three-span bridge at Trowell on the London Midland Region is a good modern example. The twin girders are of hollow trapezoidal section, which is one having important torsional resis-tance where eccentric loading occurs due to one train on one of the two tracks.

This section was adopted for the fine pre-stressed-concrete bridges at Penrith on the same Region, and it is of interest to note that a similar design for these bridges in welded steel was considered at the time, and found to be more costly.

An advantage of welded as opposed to riveted

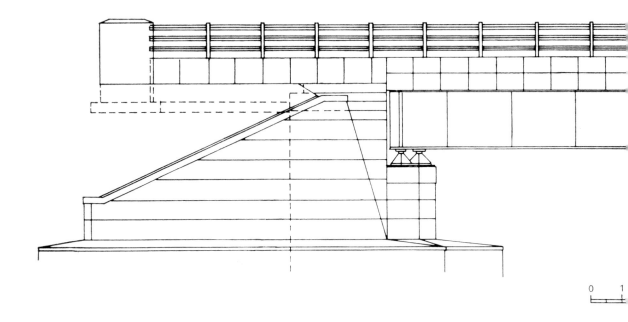

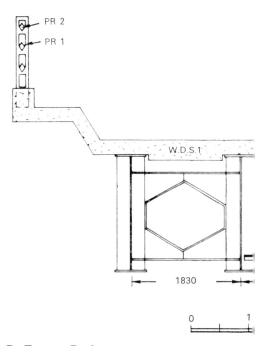

PR 2

PR 1

W.D.S.1

1830

0   1

0   1

Fig 1   *Bridge No 125. Concrete well deck on steel girders, B.R. Eastern Region*

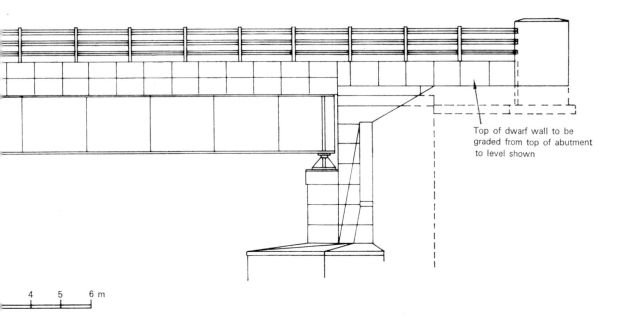

Top of dwarf wall to be
graded from top of abutment
to level shown

4   5   6 m

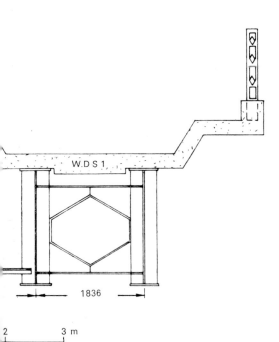

W.D S 1

1836

2   3 m

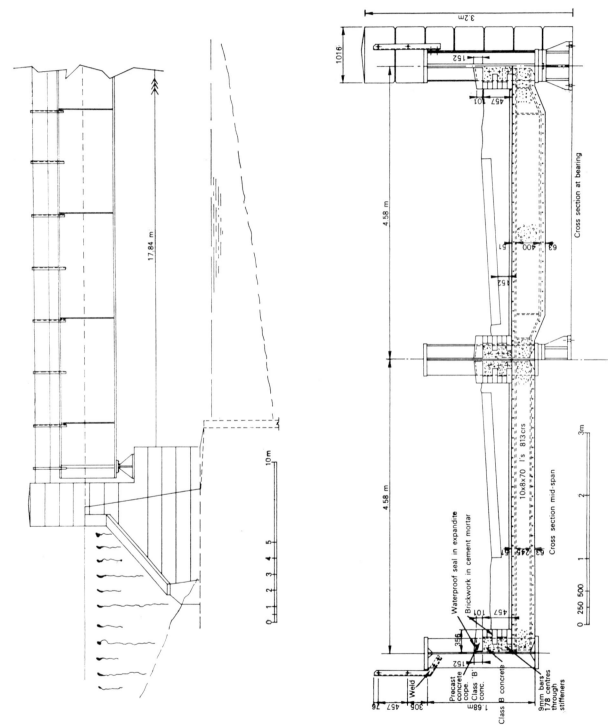

Fig 2 *Bridge No 123. Welded steel 'half-through' type, B.R. Scottish Region*

steel construction is the smooth exterior presented for painting. This, however, also emphasises the great advantage of concrete construction in that there is no maintenance repainting required at all.

Welded construction makes it possible to effect a saving in weight, and therefore in cost, particularly erection costs, but on the other hand the greater weight of concrete bridges can be an advantage because they tend to have a lower natural frequency of vibration, and consequently the tracks are less 'lively'. However, the greater weight of concrete spans becomes a disadvantage when the supporting substructures have to be designed for poor ground, and these can easily become too elaborate and costly.

A German 'half-through'-type welded steel bridge

Fig 4 shows a welded steel 'half-through' design which compares with the British Railways design in fig 3. The photograph also reproduced of a German straight-sided type also compares with the British types B, C, D, and E. The German designs were evolved with similar considerations in view, and also so that they could be rolled or lifted into position.

Prestressed concrete construction

Any account of the development of prestressed concrete techniques for bridges built by the Civil Engineering Departments of British Railways must almost of necessity start with the works made necessary by electrification on the overhead-line system.

These reconstructions began with overbridges where greater headroom for electrical clearances was required for the overhead contact line system whether on the 1,500 v dc system, as on the Pennine route between Manchester and Sheffield, or the 25 kv ac system on the London Midland Region, the Liverpool Street to Colchester and Clacton on the Eastern Region, and the Glasgow suburban lines on the Scottish Region. This latter system is now, of course, the standard for all future electrification schemes except the Southern Region, which is at present on the 600 v dc conductor rail system.

The shallower depth of the prestressed concrete beams compared with steel, strength for strength, made them at once attractive for the reconstruction of overbridges where the headroom was insufficient. Reference should be made to Mr R E Sadler's paper No 6320 before the Institution of Civil Engineers in February 1959 on this subject of bridge reconstruction on the Manchester–Sheffield–Wath line and the Colchester–Clacton line of the Eastern Region.

A number of overbridges were rebuilt on these two routes using concrete beams designed on the 'partial-prestress' system developed by Dr F W Abeles. In this system some tensile stress is permitted in the concrete itself, which results in a further saving in depth over normally designed beams where no tensile stress is allowed for.

It should be realised that every inch saved in the amount of headroom to be provided represents a very considerable saving in cost. Inverted Tee-section beams laid side by side were used for the replacement of brick arch bridges, and the intervening space filled with concrete reinforced in the normal manner with steel bars.

The first bridge in the world to be so constructed on the 'partial prestress' system was an old brick arch bridge at Buck Lane, Worsborough Dale, and fig 5, which is reproduced from Mr Sadler's paper, shows the details. Further developments of the system are to be found in the cases of two bridges at Colchester and two at Clacton-on-Sea shown in fig 6. The span depth ratio was 20:7. While on the subject of shallower depth beams, mention may be made of a system developed originally by Rendel Palmer & Tritton applied to steel joists encased in concrete. If an initial deflection is given to a steel joist during the concreting of the tension flanges and maintained until the concrete has hardened, then, on the release of the load producing the deflection, the concrete is put into compression. This enables full advantage to be taken of the modern high yield point steels so that such beams can be made

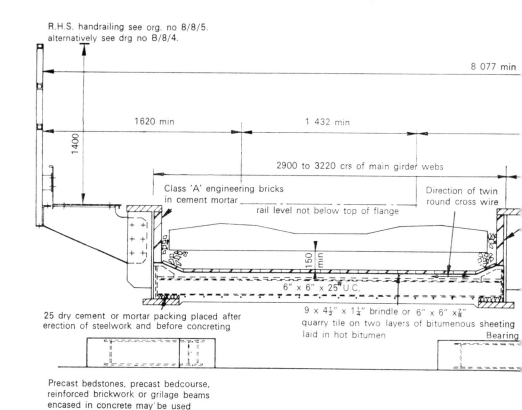

R.H.S. handrailing see org. no 8/8/5.
alternatively see drg no B/8/4.

8 077 min

1620 min

1 432 min

1400

2900 to 3220 crs of main girder webs

Class 'A' engineering bricks
in cement mortar

Direction of twin
round cross wire

rail level not below top of flange

150 min

6" x 6" x 25# U.C.

25 dry cement or mortar packing placed after
erection of steelwork and before concreting

9 x 4½" x 1¼" brindle or 6" x 6" x⅞"
quarry tile on two layers of bituminous sheeting
laid in hot bitumen

Bearing

Precast bedstones, precast bedcourse,
reinforced brickwork or grilage beams
encased in concrete may be used

Fig 3   *Standard B.R. 'half-through' type Z bridge*

50 cm   0   0.5   1m

Plate 1.3   *German 'half-through' steel bridge*

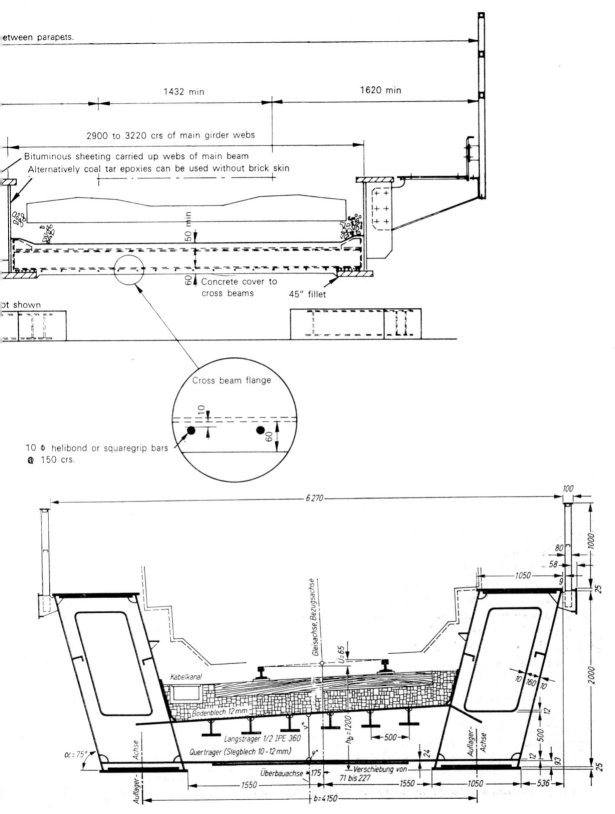

Fig 4 *German 'half-through' welded steel bridge*

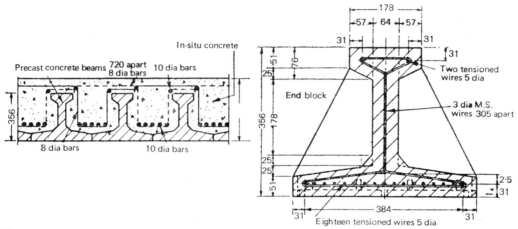

Fig 5 *Buck Lane Bridge : prestressed concrete. B.R. Eastern Region*

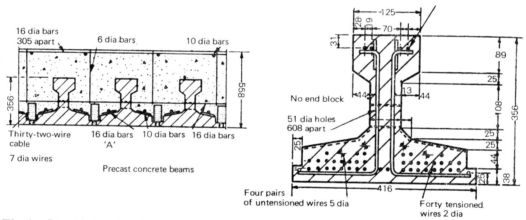

Fig 6 *Ipswich Road Bridge, Colchester : prestressed concrete*

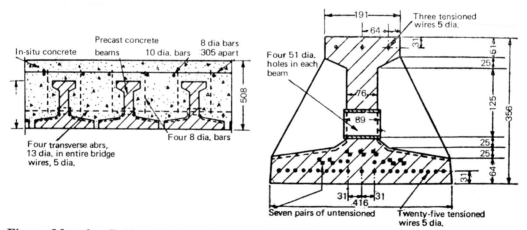

Fig 7 *Mouselow Bridge : prestressed concrete, B.R. Eastern Region*

Plate 1.4 *Bridge No. 66 Watford, B.R. London Midland Region*

shallower than would normally be the case. Also, and almost of equal importance, cracking of the encasing concrete can be avoided, thus preventing the initiation of corrosion of the steel.

With this latter end more in view, Rendel Palmer & Tritton built the Tees Dock Approach Road bridge over the Eastern Region tracks on this principle.

The system was developed and patented independently in Belgium by Monsieur A Lepski and Professor L Baes, and in this country such beams are made under licence by Boulton & Paul under the name of 'Preflex' beams. So far they have not been used by British Railways for railway underbridges. There are, however, several road overbridges on British Railways, such as the already mentioned Tees Dock Road and, for example, Scratchwood Bridge carrying the Hendon Urban Motorway over the London Midland Region main lines.

When 'Preflex' beams are placed in the bridge or other structure they are encased in concrete in the normal way over the whole surface.

As a result of the experience gained with the various alterations to bridges over electrified tracks, a series of standard prestressed concrete beams became established both on the Eastern Region, and more particularly on the London Midland Region, where they were much used on the very much greater electrification scheme between Euston and Manchester, Liverpool, and Birmingham. A photograph of the reconstruction of Bridge No 66 at Watford is typical and shows inverted Tee-section prestressed precast-concrete beams being placed, and another fine example is illustrated of Bridge No 60A at Atherstone, which was not only raised but widened for a new road. This bridge is a skew for a dual carriageway road, and sixty-six beams were used in its construction.

A further example of 'partial prestressing' is shown in fig 7, also taken from Mr Sadler's paper of Mouselow Bridge near Dinting on the Manchester–Sheffield–Wath line. In this case the additional mild steel bars between the beams, as at Buck Lane, were replaced by high strength wires located in the beams themselves.

A slight digression may perhaps be made here on the subject of bridge alterations necessary when electrification takes place. A comprehensive series of articles appeared in the *Railway Gazette* during 1963 describing the methods used by the Civil Engineers Department on the prodigious task of reconstructing, raising, or altering over six hundred bridges of all types to provide electrical clearances on the London to Manchester, Liverpool, and Birmingham electrification. At that time the amount of money allocated to the cost of the whole scheme was £160 million, and it was estimated that the total cost of bridge alterations was over £6 million. If the cost at that time of the installation of the overhead line equipment itself

Plate 1.5 *Bridge No. 60A Atherstone, B.R. London Midland Region*

Plate 1.6 *Bridge No. 37: rebuilding, B.R. London Midland Region*

is assumed to be of the order of £21 million, the additional cost of bridge alterations represents an increase of 31% over the cost of the overhead line.

One of the methods adopted in the case of brick arch bridges, of which there are a large number in the UK, was the use of controlled explosives. The upper part of the bridge was dismantled down to the last three rows of bricks, and a new reinforced concrete arch constructed on the top. The rows of old brickwork were then blown off by explosives. A photograph of Bridge No 37 on the Crewe–Stockport line illustrates this. In France a common method was to insert a new concrete crown arch into the existing structure, and an illustration of this on the Region de l'Est is given. Another method not extensively used was to raise the bridge bodily by jacking it up and building up the foundations. This was employed on very heavy bridges where circumstances made it desirable. This was done at Bridge No 77 at Garston near Liverpool, at Crewe Station, and in France at a bridge at Ars sur Moselle as illustrated. This problem of low overbridges is much more severe in Britain than in most other European countries. In no other country are there so many relatively low overbridges per route-mile. For example, in France, after the war, typical figures for bridges which required to be altered on account of electrification were 0·34 to 0·53 per route-mile.

Plate 1.7   *Rebuilt arch bridge, SNCF Region de l'Est*

In the UK the corresponding figures were from 0·9 to 2·32 per route-mile, and between Crewe and Manchester the figure was 1·89. For the whole electrified route from Euston the figure was 1·52.

In less industrialised countries level crossings are the rule, rather than overbridges, and in any case the loading gauges are almost always on a more generous scale.

All this is rather away from the subject of this book, which is about bridges carrying railways. Before, however, commencing to deal with concrete underbridges, there is one overbridge of some importance which was built by railway engineers acting as virtual subcontractors to the Ministry of Transport to carry an important

road over Kingsmoor marshalling yard near Carlisle. Many of the details of this large bridge were subsequently incorporated in later bridges which are described later.

Mossband Bridge

This large and interesting bridge was built during 1962–4 by the Civil Engineering Department of the LM Region of British Railways. It carries the Glasgow–Carlisle trunk road over the tracks of the then new Kingsmoor marshalling yard. It was designed in collaboration with the Ministry of Transport by the Bridge Section of the Chief Civil Engineer's Department. It is a very long bridge, and is set at a considerable skew. The two carriageways are structurally separate, although actually adjacent. It is 263·3 m long, and some of the design features have been repeated on subsequent bridges which have been designed and built for the national motorway programme. A diagrammatic sketch of the bridge is shown in Fig 8. Each of the two carriageways consists of four 'table' spans and four 'suspended' spans which rest between the 'table' spans.

The 'table' spans are 44·6 m long and rest on two piers 34 m apart, giving an overhang at each end of 5·3 m on which is formed the 'stepped' support for the 'suspended' span. The southern abutment provides a 'fixed' bearing for the first span which is a 'table' span. There then follow three 'table' spans with 'suspended' spans in between, all on roller bearings. The last span is a 'suspended' type with its outer end resting on a 'free' bearing on the northern abutment. The construction methods were of interest. After the completion of the mass concrete abutments and the pier foundations, the 'table' spans were built *in situ*, using steel formwork on military-type trestles. The photographs reproduced show that there were four sets of supports, two of which were subsidiaries resting on the outer plinths of each pier, and close up against the piers themselves. The function of these subsidiaries was to support the form plates over the span bearings on packings which could be withdrawn as soon as the spans were cast. The 'table' spans were

Plate 1.8 *Raised bridge, SNCF Region de l'Est*

Plate 1.9 *Formwork for 'table spans' of Mossband Bridge, B.R. London Midland Region*

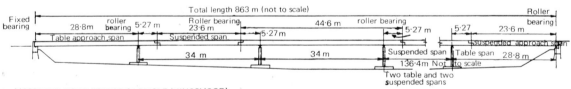

MOSSBAND ROAD BRIDGE CARLISLE (KINGSMOOR)
Prestressed concrete alternate table and suspended spans

Fig 8 *Mossband Bridge: prestressed concrete overbridge, B.R. London Midland Region*

of reinforced concrete cellular construction post tensioned. As soon as the spans were ready for the post tensioning cables to be stressed, the packings over the bearings were removed. The effect of the post stress was to impart an upward concavity to the spans, and so to transfer the whole of the weight off the main trestles, and on to the bearings on the piers. The trestles were then removed, and used on the next span to be built.

When the 'table' spans were completed the 'suspended' spans were placed. These were made at Iver by the Concrete Development Association, and brought by rail to Longton, off-loaded on to lorries, and conveyed to the site. The photograph shows how they were launched along a temporary steel girder and placed in position between each pair of 'table' spans. Each 'suspended' span consists of six precast prestressed concrete beams. After being placed, transverse stressing rods were inserted, and the space between the beams filled with concrete. After this had hardened the rods were tensioned up. The concrete deck was then cast on top. Each 'suspended' span rests on 'Glacier' bearings on the cantilevered ends of the 'table' spans. The piers supporting the 'table' spans are on roller bearings.

The foundation is a bed of sandstone at a minimum depth of 7·4 m overlaid by water-

Plate 1.10 *Completed 'table spans' of Mossband Bridge, B.R. London Midland Region*

Plate 1.11 *Placing beams for 'suspended spans' of Mossband Bridge, B.R. London Midland Region*

bearing stratum. The abutments and piers are founded on cast *in situ* tapered concrete piles.

The electrification of Euston–Liverpool–Manchester–Birmingham section of the London Midland Region not only required the alteration and reconstruction of over six hundred over-bridges, but also an extensive modernisation of the system was undertaken at the same time, and a certain number of underline bridges were reconstructed in prestressed concrete either on account of track rearrangements or to provide greater strength to cater for the more intensive traffic expected after full electric services came into operation. Examples of these conditions occur in the Manchester area at Fairfield

Street where the connections of the former Manchester South Junction and Altringham Railway with the former London Road Station were modified, and the two bridges, Stockport Road and Hyde Road which were strengthened to deal with the anticipated increased traffic from Crewe into the London Midland Region Station. Reference may be made to Mr F Turton's paper to the Institution of Civil Engineers in September 1961.

As a consequence of these reconstructions, a number of standardised prestressed concrete beam designs were created which have been used on both overline and underline bridges. On British Railways the general practice is to use prestressed or reinforced concrete slabs for bridges up to 5 m span. For spans up to 14 m standard prestressed concrete beams are laid side by side with a concrete filling cast round them to form the deck. Up to 25 m span specially designed prestressed concrete units or slabs are usual. It is now sometimes the practice with concrete bridges to attach the rails directly to the concrete deck as has already been the practice on the Netherlands Railways for some time.

As has already been mentioned, it is on the London Midland Region that the most striking designs of prestressed concrete bridges have appeared. The bridges at Besses o' th' Barn, Frodsham, and the Penrith bridges are examples, and more recently a bridge at Derby which is unique in many respects. Many of these bridges consist of single monolithic concrete beams, as at Penrith, or are built up of prefabricated sections, as at Besses o' th' Barn, and post tensioned together. Others are built up of a series of precast prestressed concrete beams laid side by side, and having a concrete deck cast on top of them. A recent bridge on the Eastern Region, No 85A, consists of two prestressed concrete slabs each carrying two tracks with the parapets cast integrally with each half-slab. The slabs were cast at the track side, and rolled into position on to previously constructed foundations and bearings.

Plate 1.12   *Completed Mossband Bridge, B.R. London Midland Region*

2 New erection methods

New foundation construction methods

The long and heavy new bridges necessitated by the new motorway system created problems of their own by reason of the paramount requirement by operating departments to reduce interference with traffic to an absolute minimum, since they had to be inserted into existing routes which were almost always busy ones. The practice, now universal, is to agree the bridge design with the Ministry of Transport, or the Local Authority where necessary, and to construct the bridge in advance of the roadworks. This makes things easier for everybody. The relatively heavier engineering work for the railway bridge can be done without the safety precautions which would have to be taken to protect the road engineers, and when the road is built, being under the railway, and with the bridge abutments and piers already built, this work becomes quite straightforward. The new bridges are almost always built alongside the tracks and then rolled or slid into position on to the already prepared bearings. This operation can be easily carried out during a long week-end shutdown which the operating departments can tolerate.

It may be said that there are three methods of constructing abutment walls and piers under the tracks while maintaining the train service, and all are in use depending on the site conditions. Examples of each type will be given when describing various different bridges in succeeding chapters.

First there is what may be called the sheet pile wall method. This consists of driving two lines of steel sheet piles transversely across the tracks opposite each pier or abutment position at the appropriate distance apart to suit the foundation width. The tops of these lines of piles are then bridged across at the top

Plate 2.1 *Stevenage Bridge No 85A : sheet piling at site, B.R. Eastern Region*

by beams capable of sustaining the weight of the tracks. The work of excavating for foundations and the construction of foundation can then be carried on under the tracks within the sheet pile walls thus formed. Line possessions are required during the pile-driving, but this can easily be arranged at night, and usually only one track at a time need be affected. The possessions are only of relatively short duration. A modern example of this is the case of Bridge No 85A on the Eastern Region of British Railways, and is shown on fig 9 together with

20

some photographs. The method was also used for the Penrith bridges described later, and others.

The second method, which is of importance under busy lines where interference with the running of trains must be kept to a minimum, is the application of 'pipe-jacking' techniques. This consists of thrusting large hollow concrete sections through the ground by powerful hydraulic jacks. The sections can then be filled with concrete and thus form a massive and stable wall or pier. The process can be carried out without traffic interruption, and without disturbing the railway formation, since, the void created by the soil removal is completely filled by the hollow concrete section. The 'pipe-jacking' system is one which has long been used for driving relatively small concrete or steel pipes for gas,

water and other services. Pipes up to 2 m dia. and in lengths of some 2 m have been used. The first pipe section is provided with a steel cutting shield, and is pushed in followed by the next and successive sections until the length of pipe is completed. The spoil is excavated from behind the shield and conveyed back along the completed pipe to the driving end. Very long lengths of pipe up to 100 m or more can be so driven. The driving gear is usually housed in a pit lined with sheet piles excavated at the driving end, the back of which forms an abutment against which the jacks can thrust. Suitable caulking rings are inserted between each successive section.

This method has been extensively used at places where laying pipes by 'cut and cover' methods would have caused too great public

Plate 2.2 *Bridge No 85A: Stevenage BR Eastern Region. Piling in centre tracks*

inconvenience, particularly under roads, and obviously under railways. In principle, it is clearly related to the method used for boring the London tubes in which a shield equipped with a cutting edge is moved forward by hydraulic jacks thrusting against a lining which is built as excavation proceeds. This was the method used to drive a tunnel through the main line embankment of the London Midland Region at Nash Mills, near King's Langley, about 1949, and was successfully completed without a speed restriction. From this, and other experiences, developed ideas for thrusting large concrete sections through suitable ground to form horizontal 'caissons' which, in cross section, could be circular or rectangular. The method has been further developed by B.R. and Tube Headings Ltd. A notable example of this is the pedestrian subway under the LMR main line at Abbots Langley near Watford, and another at Richmond on the Southern Region. The development of this process to the thrusting of large hollow rectangular section concrete segments was first tried out by the Western Region at Alphon Brook near Exeter where new river works required the insertion of a short new bridge into the WR main line. The reinforced concrete segments were 3·5 m × 2·6 m outside dimensions. They were 23 cm thick and cast in lengths of 1·2 m. It is worth describing the construction of these abutments at this point, since later developments of the system applied to larger bridges can be included in the descriptions of the bridges concerned in subsequent chapters. Fig 10 shows a plan, and a longitudinal section through one of the drives. Sheet pile-drive pits were constructed on the down side of the track, 4·6 m square and 4·6 m deep, opposite each abutment wall. A reinforced concrete thrust wall was built on the back face of the drive pits. A sheet pile wall was also built on the up side at the end of the drive. For the two drives twenty-five segments were required, twelve for one wall and thirteen for the other. A soil mechanics investigation was made before the work started to ascertain the strata to be encountered. Six 100-ton jacks were used for the driving, which was carried

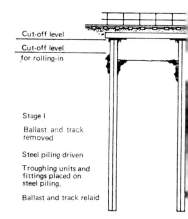

Cut-off level

Cut-off level
for rolling-in

Stage I

Ballast and track removed

Steel piling driven

Troughing units and fittings placed on steel piling.

Ballast and track relaid

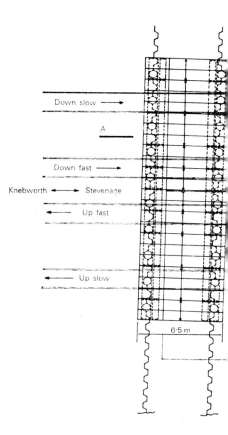

Down slow ⟶

A

Down fast ⟶

Knebworth ⟵ ⟶ Stevenage

⟵ Up fast

⟵ Up slow

6·5 m

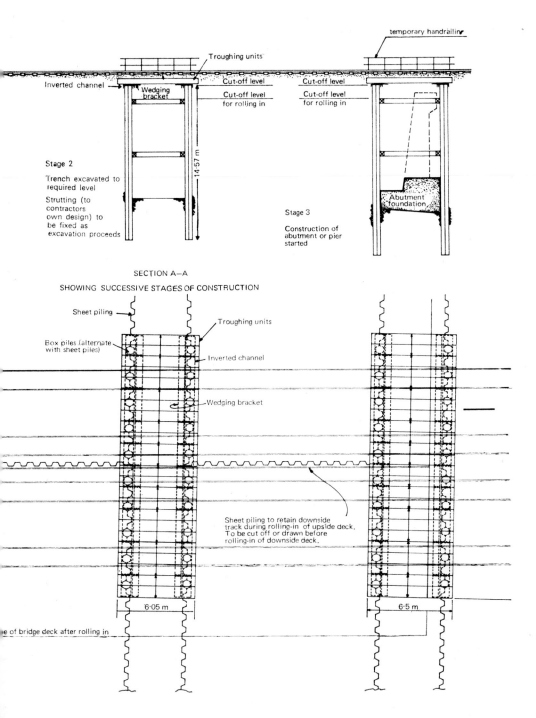

temporary handrailing

Troughing units

Inverted channel

Wedging bracket

Cut-off level

Cut-off level

Cut-off level
for rolling in

Cut-off level

Cut-off level

Cut-off level
for rolling in

14·57 m

Stage 2

Trench excavated to
required level

Strutting (to
contractors
own design) to
be fixed as
excavation proceeds

Abutment
foundation

Stage 3

Construction of
abutment or pier
started

SECTION A—A

SHOWING SUCCESSIVE STAGES OF CONSTRUCTION

Sheet piling

Troughing units

Box piles (alternate
with sheet piles)

Inverted channel

Wedging bracket

Sheet piling to retain downside
track during rolling-in of upside deck.
To be cut off or drawn before
rolling-in of downside deck.

6·05 m

6·5 m

e of bridge deck after rolling in

PLAN SHOWING LAYOUT OF TROUGHING UNITS AND PILING

Fig 9   *Bridge No 85A Stevenage: sheet piling for foundations, B.R. Eastern Region*

out under traffic conditions. As the depth below rail level was shallow, a speed restriction of 48 km/h was imposed as a safety precaution. Where greater working depths exist this restriction is often not considered necessary. A specially designed steel driving shield with a cutting edge was placed at the front of the first segment.

As the longitudinal section shows the segments were driven with the 3·5 m side upright, and the spoil was excavated at the face, loaded into skips, and conveyed back to the driving end through the hollow tunnel. The two drives occupied about two weeks. On completion, a concrete partition wall was built inside, and the half on the inside faces of each abutment wall filled with mass concrete. The other side was kept free for drainage purposes. The two walls were capped with precast concrete bed courses and bed stones to form the bridge seatings.

The bridge itself is a 26·6 m long skew, 9·8 m wide, and consists of three steel box girders with a prefabricated steel deck built in six sections, three for each track. It rests on steel bearings. It was erected by two cranes during a week-end line possession. The 'pipe-jacking' method has been applied to other larger bridges on the Western Region, and also on the Southern Region which are described subsequently. There is also a further interesting use of it on the LMR at a bridge at Derby.

The third method of building foundations is to build outside the tracks, as, for instance, at the Frodsham bridges on the LM Region. Here each bridge support rests on two reinforced concrete bored piles sunk on either side of the tracks, and subsequently bridged by a reinforced concrete beam. This beam was placed during the possession of the tracks for moving in the

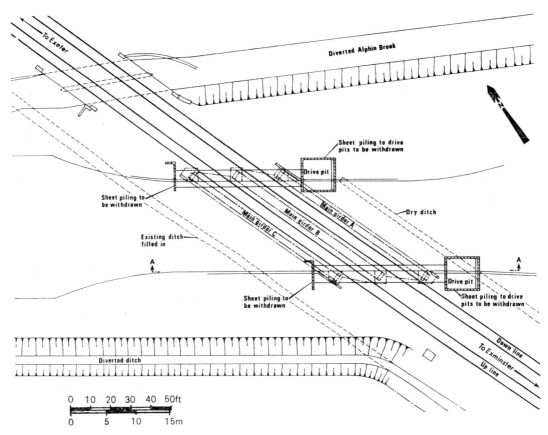

Fig 10　*Alphon Brook Bridge: 'pipe-jacked' foundations, B.R. Western Region*

Plate 2.3 *Churchdown Bridge: detail of sledges, B.R. Western Region*

surfaces involved can take higher bearing loads, which becomes important for the large and heavy bridges which are now often required. The prestressed reinforced concrete bridges such as Clifton weigh over 2,500 tons. Rolling-in is adequate for bridges up to 1,500 tons weight, heavier bridges are moved in more easily by sliding, and the possible greater cost of this method becomes justified.

Sliding was first studied in Sweden, and was applied in Germany, and also in Venezuela, but was used rather for the launching of the girders of multi-span bridges between one span and the next. In 1945 the Civil Engineers Department of the London Midland Region started a series of experiments at Crewe, making use of an old half-through-type steel girder bridge loaded up with old rails. The essential feature was the use of sledges fitted with runners of sintered bronze impregnated

bridges themselves which had been erected beside the tracks on extensions of the bearing beams, and which formed the moving-in beams. This method avoids altogether any interference with traffic until the time comes to move the bridge into position.

Moving bridges into position can, in the case of shorter steel girder bridges, be done by crane, as was done at Alphon Bridge described above. Larger bridges, especially concrete ones, have to be moved in either by rolling or by the new technique developed on the LM Region by sliding. When bridges are rolled in this is done on hardened steel balls rolling in channel sections, which are often old bull-head rails laid on their sides on temporary rolling-in beams placed opposite the abutments and piers of the new bridge. The new bridge will have been erected at the side of the tracks on the rolling-in beams.

The advantage of sliding is that the flat

Plate 2.3a *Churchdown Bridge: rolling-in completed, B.R. Western Region*

with PTFE (polytetra fluoro ethelene) sliding on rails surfaced with stainless-steel strips. The test bridge was loaded up to 800 tons, distributed so that the load at one end was approximately twice that at the other end. At the heavy end the bridge was supported on a beam provided with five sledges each fitted with twin runners of sintered bronze PTFE impregnated, sliding on two flat bottom rails with stainless-steel strips attached to the top surfaces with epoxy-resin glue. The lighter end was supported by two single-runner sledges, at the heavy end the load was distributed on to the sledges through the medium of five nests of hydraulic jacks which were interconnected so as to even out inequalities on the rail surfaces. No jacks

were used at the lighter end. The total travelling length was 10 ft (3·1 m).

The towing winches were reversible so that the bridge could be moved to and fro. It was found that the towing force varied from 18 to 37 tons at the heavy end, and from 7 to 23 tons at the light end, and the time required for the 10 ft movement varied from 7 to 34 minutes.

From these tests was developed the present sliding-in arrangements, and the first full-scale trial was made early in 1967 on a bridge at Rowley Regis with a 50 m long prefabricated concrete skew slab weighing 1,500 tons. It was moved in in a net time of 15 minutes. This was a rehearsal for the much larger and heavier

Plate 2.4a *Sliding-in trials at Crewe, B.R. London Midland Region*

Plate 2.4b *Sliding-in trials at Crewe, B.R. London Midland Region*

Clifton Bridge near Penrith, moved in later in the same year. The operation was controlled by closed-circuit television. Attachment of the stainless-steel strips to the rails is now by studs, epoxy resin having proved unreliable. Mild steel billets are normally used instead of old rails, and a strip width of 76 mm is found to be sufficient. The next bridge to be moved in by sliding was the Clifton Bridge, one of the three Penrith bridges described later. It was a three-span prestressed concrete monolithic beam 74 m long and weighing some 2,500 tons. Fig 11 shows details of one of the sledges used. They consist essentially of a 13 mm thick towing plate 72 cm wide, and 3·8 m long, with six pairs of slots 8·9 cm wide and 51 cm long

into which are inserted the PTFE runners. There is a 38 mm thick vertical towing plate welded on at the towing end which is secured to a built-up welded steel supporting bracket by an 89 mm dia steel pin. The bracket is secured to the concrete bridge girder by eight 25 mm dia Rawl bolts. Along the edges of the towing plate there are steel guide angles, and there is a substantial end plate welded on. At the piers of the bridge there were six packs of steel plates which incorporated two Freyssinet flat jacks in each pack. These jacks not only serve to raise and lower the bridge girder on to its final seatings but are hydraulically interconnected so as to equalise the pressure on the running-in surfaces where inequalities might exist. The

Plate 2.4c *Sliding-in trials at Crewe, B.R. London Midland Region*

sliding-in rails were 20 cm×7·6 cm, and the stainless-steel strips were No 8 gauge. Strips were also provided along the sides of the rails. The distance to be moved in the case of Clifton Bridge was 18·6 m, and the whole operation was successfully completed during a 54-hour week-end possession of the two tracks.

Since then the sliding-in method has been used on a number of other bridges of comparable weight.

Although it is now almost always the practice to build new bridges alongside the tracks, and to roll or slide them in, it is sometimes possible or even desirable to either close the railway or divert it. At Fairfield Street Bridge, Manchester, conditions made it possible to close the station

over the street and to terminate trains at the previous station without serious operational inconvenience. At Besses o' th' Barn Bridge on the Manchester–Bury line the two tracks were singled into a diversionary track round the bridge site. The site conditions existing here, and the type of bridge which was required made this practically a necessity.

Where there are four tracks, temporary diversion of the traffic on to two out of the four is often possible without undue interruption to the train services, and this often simplifies the civil engineering work. As the bridge for a four-track route is often made in two halves, the moving-in operation takes place over two separate week-end possessions, two

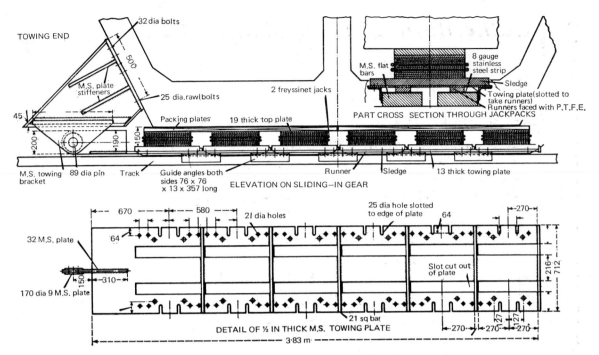

TOWING END

32 dia bolts

M.S. plate stiffeners

25 dia. rawl bolts

45

M.S. towing bracket

89 dia pin

Track

Packing plates

Guide angles both sides 76 x 76 x 13 x 357 long

2 freyssinet jacks

19 thick top plate

Runner

Sledge

13 thick towing plate

M.S. flat bars

8 gauge stainless steel strip

Sledge

Towing plate (slotted to take runners)

Runners faced with P.T.F.E.

PART CROSS SECTION THROUGH JACKPACKS

ELEVATION ON SLIDING—IN GEAR

670   580   21 dia holes

25 dia hole slotted to edge of plate   64   270

32 M.S. plate   64

170 dia 9 M.S. plate   310

Slot cut out of plate

216
712

21 sq bar

DETAIL OF ½ IN THICK M.S. TOWING PLATE   270   270   270

3·83 m

Fig 11   *Sliding-in sledge for Clifton Bridge, B.R. London Midland*

tracks being closed at the time while traffic proceeds on the other two tracks. Bridge No 85A at Stevenage is an example.

While these problems continually arise in these overcrowded islands they are not so frequent on overseas railways. In European and other countries diversionary tracks are usually much more easily arranged than in England. In any case, the work of the railway engineer is often less restricted by conflicting authorities than is the case here, and he often enjoys greater freedom of action.

3 *Some notable steel bridges in the UK*

Grosvenor Bridge (Southern Region)

The reconstruction of this important bridge over the River Thames is one of the major bridge works undertaken by British Railways since the war. Not only is it a fine example of modern welding technique, but the conditions under which it was rebuilt called for the most careful and elaborate planning and organisation. Formerly known as Victoria Bridge, it consisted of three separate bridges built at different times. The first two-track bridge was built in 1859, and soon afterwards in 1865 it was widened on the downstream side to take five more tracks. In 1901 it was again widened on the upstream side for a further two tracks, making nine in all. It was also strengthened from time to time as the traffic increased annually to a degree never contemplated at the time it was built. The cost of maintenance, however, became increasingly uneconomic, and in 1958 Messrs Freeman Fox & Partners were asked to prepare a report on the practicability of constructing an entirely new bridge. The conditions laid down by British Railways were severe. The new bridge was to carry ten tracks, and eight had to be maintained in traffic during the reconstruction. The Port of London Authority also stipulated that out of the four river spans two were to be kept open to navigation. The two road spans at each end were also to be kept open to traffic along Grosvenor Road and Spicers Wharf, and allowance for the widening of Grosvenor Road had to be made. At the time of the report the bridge was carrying over a thousand trains daily. Fig 12 shows the outlines of the old and new bridges.

Both the old and the new bridges are the same length, 282 m, and consist of four river spans and two land spans. Although the old bridge had arches of the same shape, each of its portions differed, particularly in the matter of the foundations. The superstructures were in wrought iron except the 1901 bridge, where the material was steel.

The first bridge was built on a concrete raft

Plate 3.1   *Grosvenor Bridge rebuilding, B.R. Southern Region*

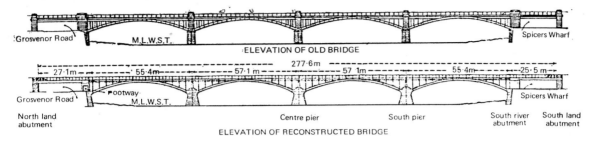

Fig 12 *Grosvenor Bridge. Old and new bridges, B.R. Southern Region*

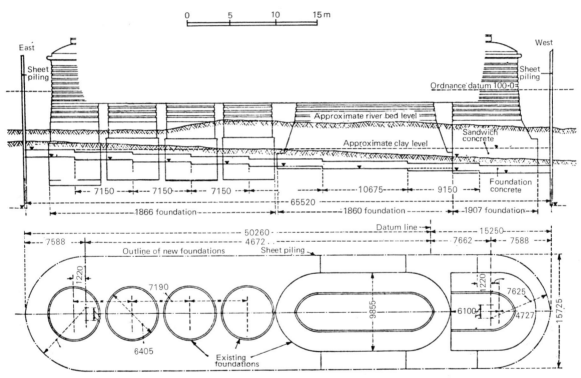

Fig 13 *Grosvenor Bridge reconstruction: foundation pier plan, B.R. Southern Region*

foundation with brick piers faced with lime-stone carrying four two-pin arches in wrought iron. The second was founded on four cast-iron caissons 6·4 m dia, sunk into the river bed and filled with concrete. On these were built brick piers, and the deck stringers were continuous over the piers and abutments forming with the wrought-iron ribs a contin-uous girder of varying depth. The last addition

was founded on a single large cast-iron caisson filled with concrete, with brick piers carrying steel arch ribs, the tracks carried on deck plates with steel ballast walls. A plan and eleva-tion of the four foundation types is shown in fig 13. The piers of the two last bridges were granite-faced. Although the overall length of the new bridge is identical with the old, the four river spans are shorter—51 m instead of

54 m—due to the necessity of increasing the width of the piers and strengthening the substructures to deal adequately with the specified increase in loading required. The new bridge has been designed to carry twenty units of RAI BS 153 loading, which is approximately twice that for the original bridge.

Phase 1 of the reconstruction was therefore the reconstruction of old piers and abutments within sheet pile coffer dams protected by substantial dolphins against the powerful river currents. The footings were enlarged from 9·6 m × 61 m to 14 m × 67 m and fig 14 is a part section of the new pier showing how the old pier was stripped down to the brick core and the new concrete keyed into it. This work had to be done step by step according to the order in which the existing tracks could be given up by the operating department and commenced with track No 9 on the extreme downstream

side. Masonry demolished in the reconstruction of the piers amounted to 6,930 m³, and this was replaced by 19,030 m³, of new concrete and 400 tons of steel reinforcing bars.

Messrs Freeman Fox's report was accepted and design work was completed for work to commence in early 1963. It was finally completed in 1968.

For the dismantling of the old arches and the erection of the new a lattice steel service girder was designed and assembled on a nearby siding made available by the Southern Region. This girder was made sufficiently long to bridge between two piers, and strong enough to lift off the old arches on to pontoons and replace them with the new ones. The service girder when required was traversed on to the track involved and moved into position on special bogies over the span which was to be dismantled. The old arches were disconnected

Plate 3.2   *Grosvenor Bridge rebuilding: river works, B.R. Southern Region*

Plate 3.3   *Grosvenor Bridge rebuilding : floating away old arch ribs, B.R. Southern Region*

Plate 3.4   *Grosvenor Bridge rebuilding : erecting new ribs by service girder, B.R. Southern Region*

and lifted off on to the pontoon and floated downstream on a falling tide. Similarly the new ones were floated in on a rising tide and lifted off the pontoon by the service girder. A photograph of the operation is reproduced.

The new arches, fig 15, are composed of two welded box-section ribs, 62 cm wide and 114 cm deep, and are built up of flange plates 31 mm thick, except for the middle 8·5 m where the thickness is 22 mm. The steel battle deck is supported on the ribs by thirty-two tubular-steel struts 17 cm dia. The longest ones at the ends are fixed top and bottom, while the intermediate ones have a hinged joint at the bottom. A cross-section through the deck and arch is shown in fig 16. Suitably disposed longitudinal and transverse stiffening is provided on both ribs and deck.

The ribs were fabricated by Dorman Long & Co, and the Teeside Bridge & Engineering Co. They were assembled at a site on Nine Elms Wharf in half-span units which were loaded on to pontoons and floated upstream on the rising tide. Each unit was complete with deck, walkways, cable ducts, drainpipes, etc. On arrival at the bridge site they were lifted off by the service girder and were held suspended while the two halves were bolted and welded together, and then lowered on to jacks located

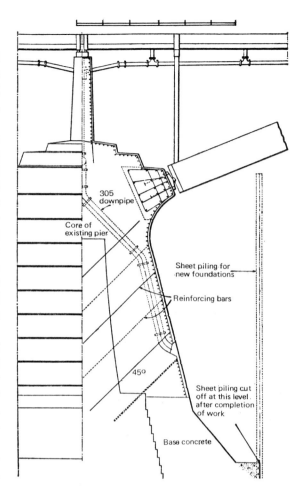

Fig 14 *Grosvenor Bridge reconstruction: elevation of new pier, B.R. Southern Region*

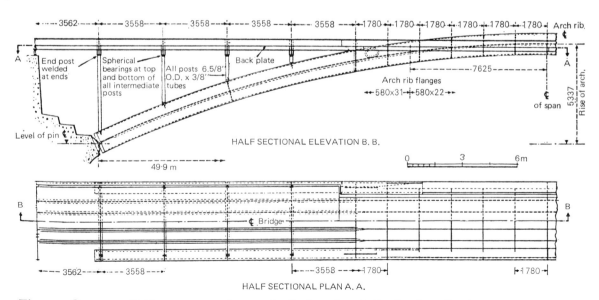

Fig 15 *Grosvenor Bridge reconstruction: arch rib elevation, B.R. Southern Region*

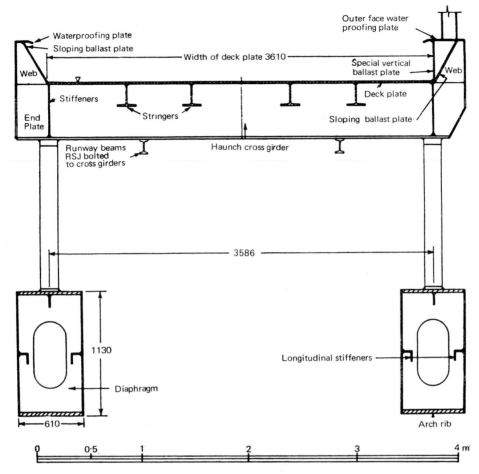

Fig 16 *Grosvenor Bridge reconstruction : arch rib cross-section, B.R. Southern Region*

at the bearings. They were thus held while the pin-joint bearings were accurately positioned and concreted in. The jacks were then released after the concrete had set.

Before fabrication of the arches was ordered, tests on Perspex scale models of the arch and deck were carried out at the Imperial College of Science. Soil investigations were made at the Soil Mechanics Department of Richard Costain Ltd, and tests on river protection for concrete surfaces at the Imperial College of Science.

Messrs Marples Ridgeway Ltd were the main contractors, and the Chief Civil Engineer for the Southern Region of British Railways at the time was Mr A H Cantrell, under whose overall supervision the work was done.

Three span bridge at Trowell (LM Region)

This is an interesting example of a combination of welded steel and concrete construction located near Trowell on the Radford–Trowell branch of the London Midland Region. An outline diagram and a cross-section of the bridge is shown in fig 17. It was built to permit of the passage underneath of the Lancashire–Yorkshire Motorway and was constructed in advance of the roadworks.

The centre span is 43 m, with two 18 m side spans, the deck is 8·5 m wide, and the clear headroom over the road is 5·5 m. The interesting feature is the use of hollow trapezoidal section-welded steel-plate girders, one for each track, which are continuous over the abutments and

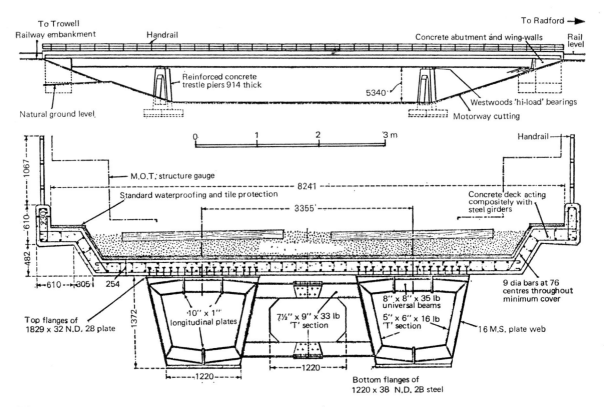

To Trowell

Railway embankment

Handrail

To Radford →

Concrete abutment and wing-walls

Rail level

Reinforced concrete trestle piers 914 thick

5340'

Natural ground level.

Westwoods 'hi-load' bearings

Motorway cutting

1067

M.O.T. structure gauge

Standard waterproofing and tile protection

8241

Handrail

Concrete deck acting compositely with steel girders

610

482

230

3355'

610 305 254

9 dia bars at 76 centres throughout minimum cover

Top flanges of 1829 x 32 N.D. 28 plate

1372

10" x 1" longitudinal plates

7½" x 9" x 33 lb 'T' section

8" x 8" x 35 lb universal beams

5" x 6" x 16 lb 'T' section

16 M.S. plate web

1220

1220

Bottom flanges of 1220 x 38 N.D. 2B steel

Fig 17   Trowell Bridge : cross-section, B.R. London Midland Region

piers. The cross-section shows the construction. The top flange plates are 1·9 m wide and 48 mm thick; the bottom flange plates are 1·25 m wide and 48 mm thick with side plates 16 mm thick, and the beam has a depth of 1·4 m. Two longitudinal flat stiffening bars 25 cm × 25 mm thick are welded to the whole length of underside of the top flange and one similar bar to the top side of the bottom flange, and at intervals of 4·8 m there is a lateral stiffening system of universal beams 20 cm × 20 cm welded to the top and bottom flange plates with Tee-sections 12·7 cm × 15 cm welded to the side plates. At the same intervals—4·8 m—the two girders are cross-connected by 18 cm × 23 cm Tee-sections with plate stiffeners 12·7 mm thick. The depth of the girders—1·4 m—is less than would have been the case had standard RSJ sections been used, and the good torsional stiffness of the section reduces the stresses in the concrete

deck when only one track is loaded. The 25 cm thick reinforced concrete deck was cast on to the girders at site and it was bonded to the tops of the girders by 3,500 steel shear studs welded to the top flanges of the two girders.

The bridge is carried on mass concrete abutments, and two reinforced-concrete portal frame trestle piers. As these piers were built before the motorway, the two legs were constructed inside two pits excavated down to the required depth. The bridge rests on Westwood HI steel roller bearings on each pier, with a load-bearing capacity of 600 tons. At the abutments the girders are encased in a concrete counterweight which rests on a bankseat formed on the abutment itself. This bridge was completed in 1965 under the overall supervision of Mr W F Beatty, the Chief Civil Engineer, London Midland Region, British Railways, and was designed and erected by his staff.

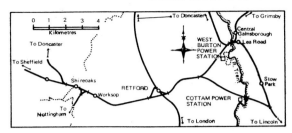

Fig 18   *Retford dive-under bridge : site plan, B.R. Eastern Region*

EASTERN REGION

The Retford 'dive-under'

All railway enthusiasts familiar with the East Coast main line to Scotland will remember the nearly right-angle level crossing south of Retford Station where the old Great Central line from Sheffield to Gainsborough and Lincoln crossed the King's Cross to Edinburgh main line.

This crossing was eliminated in 1965 by the construction of a cutting and a new bridge to enable the Sheffield to Lincoln line to pass underneath. This was part of the policy of introducing so-called 'merry-go-round' trains to supply the new power stations being built along the River Trent by the Central Electricity Generating Board, particularly those at West Burton and Cottam, with coal direct from the south Yorkshire collieries. The small site plan, fig 18, shows the scheme of which the 'dive-under' is part, consisting of relaying the former GCR line in a cutting between Bridge No 192 and Thrumpton Lane, making a new connection to the north loop into Retford Station, and constructing a new south loop to rejoin the new diversion at Thrumpton Lane.

The new bridge is of interest as an example of a double-track welded half-through-section steel bridge of which the plan and cross-section are shown in fig 19. It is 12 m long on the skew and 11 m wide, including a footwalk. The really interesting feature was the method of supporting the main-line tracks while the abutments, which also form the walls of the cutting in which the new diversion is laid, were built. To enable this to be done the main-line tracks were supported on heavy way beams built up of heavy steel joists resting on three groups of piles spaced 12 m apart. The two outer groups consisted of two piles, and the middle group of five. They were capped with steel joists encased in concrete on which the way beams rested. Between the pile groups excavations for the bridge abutments were made and the abutments cast. Fig 20 shows a plan and cross-section of the works. The piles were driven during Sunday possessions of the lines and as soon as the way beams had been placed and the tracks laid traffic proceeded while the construction work was being carried out. Previously a speed restriction of 104 km/h over the level crossing had been imposed. This has now been raised to 130 km/h.

The bridge was erected by the side of the main lines and rolled into position during a Sunday possession of both tracks. During the construction period while the abutments were being built a speed restriction was imposed. Before being rolled in, the cross girders forming the bridge floor were encased in concrete with steel fabric reinforcement. The top surface of this deck was finished off with blue tiles bedded in with boiling bitumen

The necessity for the elimination of the old crossing at track level will be appreciated by the fact that for one power station alone an annual traffic of coal of $5\frac{1}{2}$ million tons is involved. The 'merry-go-round' system involves continuous movement of the coal trains which consist of special wagons which can unload the coal automatically at the power-station sidings while the train is in motion so that it immediately returns empty back to the loading depots at the collieries.

Mr A K Terris was the Chief Civil Engineer of the Eastern Region at the time this work was undertaken. The general contractors were Henry Boot and Sons (Civil Engineering Ltd) and the sub contractor for the steel work was the Butterley Engineering Co.

Welded plate girder bridge at Tinsley (Eastern Region). Single span of 52 metres

This bridge over the Sheffield and South Yorkshire Canal, at the west end of Tinsley

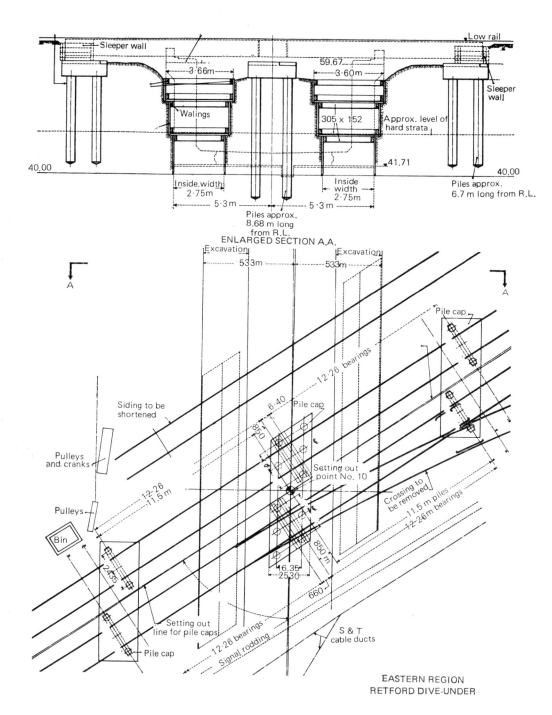

Sleeper wall

Low rail

3·66m

59.67

3·60m

Sleeper wall

Walings

305 × 1·52

Approx. level of hard strata

40.00

41.71

40.00

Inside width
2·75m

Inside width
2·75m

Piles approx.
6.7 m long from R.L.

5·3 m

5·3 m

Piles approx.
8.68 m long
from R.L.

ENLARGED SECTION A.A.

Excavation

Excavation

533m

533m

A

A

Pile cap

12·26 bearings

Siding to be
shortened

6·40

Pile cap

850

Pulleys
and cranks

Setting out
point No. 10

Pulleys

12·26
11.5 m

Crossing to
be removed

11.5 m piles

12·26 m bearings

Bin

850 m

2455

16.35

2530

660

Setting out
line for pile caps

12·26 bearings

Pile cap

Signal rodding

S & T
cable ducts

EASTERN REGION
RETFORD DIVE-UNDER

Fig 19   *Retford dive-under bridge: plan and cross-section, B.R. Eastern Region*

Plate 3.5 *Trowell Bridge, B.R. London Midland Region*

marshalling yard, is one of the largest single-span welded plate girders in Britain, with a clear span of 52 m, and weighs 440 tons. Each girder is 4·3 m deep and was fabricated in three sections for ease of erection; there are two end sections and one centre section, making six sections for the two side girders.

The sections are made in two halves longitudinally. The top flange consist of two 5 cm thick plates welded together along the sides, one is 91 cm wide and the top plate 86 cm wide. To this flange plate is welded a web plate 19 mm thick and 1·6 m deep, and along the lower edge of this web is welded a Tee-section 46 cm × 42 cm. This section is actually half a standard universal beam 92 cm × 42 cm × 253 kg, cut longitudinally. This forms a member with a total depth of 2·15 m, and two such sections bolted together with HSFG (high strength friction grip) bolts forms the complete girder 4·3 m deep. These girders are stiffened on the outside by vertical Tee-section members at intervals of 3·2 m, and the members themselves are built up from a 46 cm plate 38 mm thick welded to a web 38 cm × 25 mm thick, the whole being welded to the main 19 mm web plate. The main constructional details are shown in fig 20. On the detail of the outside stiffeners should be noted that on the bottom half the lower end of the vertical member is turned in

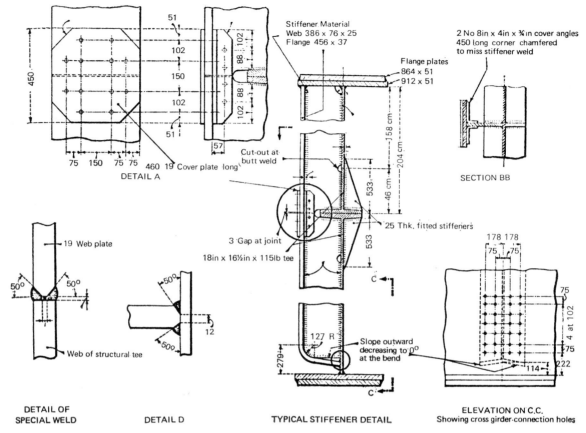

Fig 20 *Tinsley Canal bridge : details, B.R. Eastern Region*

so as to leave a gap between it and the bottom flange of the main flange plate. As this member is in tension, and susceptible to the effects of fatigue, this construction leaves it free. The only additional stiffening on the inside of these girders are the 25 mm thick triangular plates welded opposite each outside stiffener. The completed bridge showing the inside and outside stiffeners is illustrated.

The end sections of the bridge are built up in the same way but with suitable modifications to arrange for the seatings. The bearings are pin-jointed with some longitudinal movement allowed at one end.

The floor is of battle deck construction built up of 25 mm thick plates supported on universal Tee-stringers with fabricated Tee cross-girders bolted to the two main welded girders. Some 12,000 HSFG bolts were used in the construction.

The contractor was Henry Boot & Sons (Railway Engineering) Ltd and the Butterley Engineering Co Ltd were sub-contractors for the steel work. The whole work was under the overall supervision of the then Chief Civil Engineer, Eastern Region, British Railways, Mr A K Terris.

Vertical lift bridge at Deptford Creek (Southern Region)

This new vertical lift bridge was erected in 1963–4 to replace the old wooden bascule bridge over Deptford Creek originally built in 1838 on the North Kent line of the Southern Region.

Deptford Creek is crossed by two spans, on

the London side by a masonry arch to a pier in midstream, and on the Greenwich side by the new lift bridge from the central pier to the land abutment. The old timber drawbridge was in two halves, raised by hand winches. The operation involved lifting the track rails by eighteen men, and when a ship required to enter, a total delay to traffic amounted to $1\frac{3}{4}$ to $2\frac{1}{2}$ hours. It is astonishing that this sort of traffic interruption was allowed to continue for so long.

The new bridge is electrically operated by one man and the lifting occupies five minutes up and down. With normal train intervals of twelve to fifteen minutes it is now possible to arrange the lifting operation to take place during a normal train interval with the virtual elimination of traffic delay.

The new bridge consists of four lattice steel towers 19·7 m high which measure 2·3 m × 2 m at the top and 2·6 m × 2 m at the base. The inner faces are vertical and carry the guide rails for the lifting deck. They weigh 56 tons each. The steel lifting deck weighs 40 tons, and the lift of 12 m gives a clearance of 17 m above the highest tide level. Fig 21 shows the elevation and plan of the bridge. The towers are of hollow lattice construction and are secured to their foundations by four 38 mm dia bolts. They are connected at their tops by a lattice steel framework 1·85 m deep, on which is mounted the electric winding mechanism. Balance weights

Plate 3.6   *Retford dive-under bridge, B.R. Eastern Region*

Plate 3.7 *Tinsley Canal Bridge, B.R. Eastern Region*

totalling 37 tons for the lifting floor move in vertical guides up and down the inside of the towers.

The lifting floor, constructed of plate girders, was prefabricated in convenient sections and erected at the tower tops clear of the traffic beneath. It was site welded and bolted together using high strength friction grip bolts for the main connections.

The electric operating gear comprises two squirrel-cage 40 hp motors driving the two winches through a 50:1 reduction gear and cardan shafts, with a secondary reduction gear of 4:1.

The haulage ropes are in duplicate and consist of 6/37/15 mm stranded steel ropes of 1730/1880 kg/cm² ultimate strength.

The motors and controls were supplied by Lawrence Scott Electromotors Ltd. One motor is a standby unit, since only one is required to operate the bridge.

The existing midstream pier required little alteration, and the holding down bolts for the towers were grouted in to the existing masonry. New foundations in reinforced concrete were required on the Greenwich side which were taken down to the level of the foundations of the railway viaduct.

The sections of the towers were erected by a crane with a jib 37 m long during night possessions of the tracks. During day work on the foundations, the jib was not permitted to swing over the tracks. Mr A H Cantrell was the

Chief Civil Engineer of the Southern Region of British Railways at the time of building. The bridge was designed and the construction work supervised by the Consulting Engineers, Messrs W S Atkins & Partners, under the overall supervision of Mr Cantrell. The contractors were Sir William Arrol & Co Ltd.

Three bridges on the Western Region

The three bridges, Churchdown, Haresfield, and Hyde Lane on the Western Region have one feature in common, that the foundations have been constructed by the 'pipe-jacking' method which originated on the Western Region. This method has now become quite common and has been referred to elsewhere in this book, its

chief advantage being that it enables foundation work under the tracks to be done under traffic conditions without a shut down of the train service. Apart from this the bridges show some interesting features as welded steel structures. They were all built under the overall supervision of Mr F R L Barnwell the Chief Civil Engineer Western Region.

Churchdown Bridge

This bridge between Cheltenham and Gloucester is located at a point where the line is on an embankment some 9 m high and a bridge was required to be inserted to allow of the construction of a new motorway passing underneath the tracks. As is now invariably the

Plate 3.8 *Deptford Creek vertical lift bridge, B.R. Southern Region*

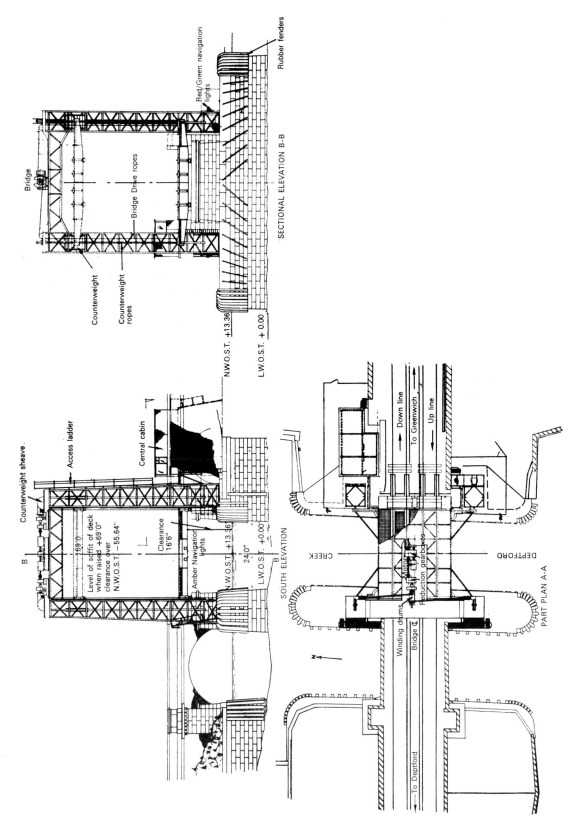

Fig 21 *Deptford Creek lifting bridge : elevation and plan, B.R. Southern Region*

Plate 3.9 *Churchdown Bridge, B.R. Western Region*

case the bridge was built in advance of the road-way.

It consists of two welded steel box-section girders one for each of the two tracks supported at each end on a welded steel plate, splayed leg, portal frame. The arrangement is shown diagrammatically in fig 22. It is a single-span structure 39 m between bearings, and the twin girders are 4·8 m apart. The centres of the legs of the portal frames are 17 m at the base and 13 m at the top. They are thus wide enough apart to allow the foundations to be constructed clear of the tracks without interfering with traffic. In order to enable the cross-member of the portal to be passed through the bank after the legs had been erected on their bases, a steel tube 2·4 m dia was 'pipe-jacked' through the bank 2·1 m below rail level from one side to the other through which the cross-member was threaded again without interference to traffic.

This tube was made from 6 mm thick steel

plates in sections 1 m long bolted together so that it could be easily dismantled when the time came to roll the bridge into position.

The constructional procedure was to build the foundations for the portal structures, and two pits lined with sheet steel piles 7·9 m × 5·9 m opposite each portal position and on either side of the tracks were excavated, and the concrete foundations placed. The legs of the portals, which were fabricated at the maker's works, were then brought to site and lowered on to the foundations to which they were secured by ten bolts 57 mm dia.

Next the two cross-members were brought to site, and by means of two cranes were thread-ed through the tubes during a night possession of the line. They were then welded on to the tops of the splayed legs. Rolling-in beams were then constructed with one end resting on the ends of the portals and the other on temporary bored piles. The box girders, fabricated at the maker's works, were then brought to site, and

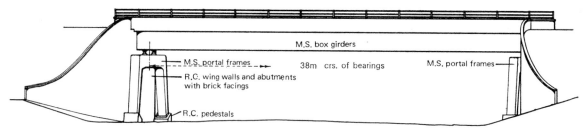

Fig 22 *Churchdown Bridge: outline diagram, B.R. Western Region*

Plate 3.10 *Churchdown Bridge: excavation for portal frame legs, B.R. Western Region*

laid on the rolling-in beams. The rolling-in rails consisted of bull-head rails lying on their sides with hard steel balls rolling in the channel so formed.

The ground was then excavated and the tubes surrounding the cross-members of the portal frames removed and the girders moved into place. Each bridge girder weighed eighty-one tons, and the rolling-in operation was carried out during the Easter week-end in April 1968,

for which full possession of both tracks was given. Closed-circuit television was used to control the operation, which, included the removal of over 1,540 m³ of the top 3·1 m of earth by four traxcavators and a bulldozer.

The general particulars of the bridge girders are: depth 2·32 m, width 1·59 m. The top plate is 41 mm thick, and the bottom plate is 47 mm thick with a stiffening plate 0·9 m wide and 31 mm thick. The side plates are 12 mm thick, and the box is braced internally at intervals with cross-frames and angle stiffeners, and the ends are sealed with plates 12 mm thick, so that being hermetically sealed no corrosion is expected. The exterior and interior were shot-blasted and given two coats of priming and iron-ore paint at the maker's works before being sealed up. They each rest on four knuckle bearings, one pair at one end being fixed, while at the other end longitudinal sliding is allowed for.

The gap between the ends of the girders and the bank top is bridged at each end by seven precast concrete Type B bridging units. During construction these permitted further work on the wing walls and abutments to be carried out behind the portal frames. Although the two girders are separate and at slightly different levels they are connected at their ends by braced steel frames. The walkways at each side were bolted on after erection, and the deck was treated with a 12 mm thick layer of glass-fibre reinforced Flintcote.

Haresfield Bridge

This bridge, located on the Gloucester–Bristol line of the Western Region is about midway

Plate 3.11   *Churchdown Bridge: 'pipe-jacked' tube for cross-member, B.R. Western Region*

between Gloucester and Stonehouse and was required to be inserted into the double-track line to allow for the M5 motorway to pass underneath. The line is on an embankment some 9·3 m high and the headroom over the road is 5 m. It is in two spans and is 46·5 m between abutment bearings, with a central pier. The road crosses on a 60° skew, and each track is supported on twin heavy steel girders with a reinforced concrete deck between them. The bridge is noteworthy in that it is the second bridge to be provided with abutments constructed by 'pipe-jacking' methods using large hollow rectangular section precast concrete segments built in three tiers. The lowest is laid on its side, and is 4·3 m × 2·17 m, and forms the base. On the inner end of this are placed the two upper sections, each 2·5 m × 1·8 m, one on top of the other, thus forming an L-shaped section abutment. These were tied together internally by stressed cables and filled with concrete, thus forming a massive and stable abutment wall. The diagram, fig 23, makes it clear. These concrete sections were driven through the bank by powerful hydraulic

Plate 3.12   *Haresfield Bridge, B.R. Western Region*

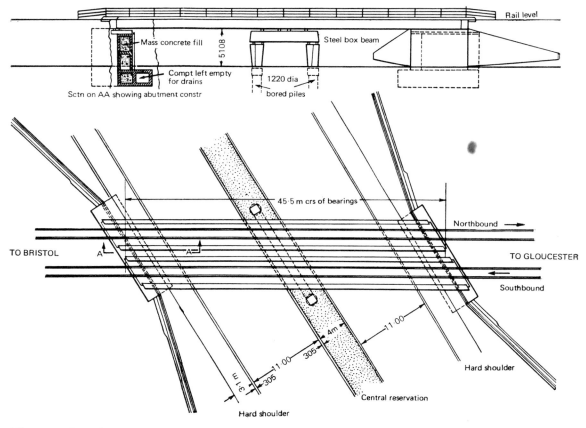

Rail level

Mass concrete fill

5108

Steel box beam

Compt left empty
for drains

1220 dia
bored piles

Sctn on AA showing abutment constr

45·5 m crs of bearings

Northbound →

TO BRISTOL

A

A

TO GLOUCESTER

Southbound ←

Hard shoulder

11·00

4m

305

11·00

305

3·1 m

305

Hard shoulder

Central reservation

Fig 23 *Haresfield Bridge: outline diagram, B.R. Western Region*

jacks. Messrs Tube Headings were responsible for this work. The central pier was formed by a heavy steel beam resting on two 1·24 m dia bored piles sunk outside the railway formation, and a steel tube was jacked through the bank at the correct height to enable this beam to be inserted without interference with traffic in exactly the same way as was described for Churchdown Bridge. The two bridge girders were fabricated at the makers works and brought to site where the concrete deck was cast on. Rolling-in beams were constructed and they were rolled into position during a week-end possession of both tracks.

To accommodate the jacking gear, and to provide a solid backing for the thrust of the jacks, sheet pile pits were excavated opposite each abutment position. The Turriff

Construction Co Ltd were the main contractors.

Hyde Lane

This bridge is at Swindon Village near Cheltenham and was the reconstruction of an existing bridge in the Birmingham–Cheltenham line of the Western Region. It was necessary to widen the bridge for road reconstruction. It is another example of the use of 'pipe-jacking'. It is a relatively short double-track bridge and was fabricated in a complete unit, and lifted in by crane as the photograph reproduced shows. The foundations consist of hollow rectangular reinforced concrete units thrust through the embankment by the methods already described.

The contractors were Johnston Bros (Shropshire) Ltd, Pipe Division, and the 'pipe-jacking' was done by Hadsphaltic Construction Ltd, Pipejack Division.

Fledborough Viaduct over the River Trent

It will perhaps have become evident to readers that the methods of erection of bridges are often more interesting than the design of the bridges themselves. In Chapter 7 there is described the Attock Bridge over the River Indus in Pakistan, where the new bridge, which was an enlargement, was erected round the old one which was used to support the erecting cranes and other gear. In the case of Fledborough the procedure was reversed, and the new bridge was built inside the old one. Both bridges were, incidentally, designed by Rendel Palmer & Tritton.

Built in 1896, Fledborough Viaduct over the Trent Valley near Clifton carried the old railway between Chesterfield and Lincoln. It is shown in fig 24. It is some 0·8 km long and consisted of four steel lattice girder spans of 36·6 m each with brick arch viaduct approaches at either end. The girders rested on twin piers of hollow cast-iron columns 3·1 m dia, filled with concrete, and braced together with steel channel and tubular sections. This bracing was badly corroded, and had to be renewed and some minor cracks in the cast-iron columns repaired. Otherwise these existing columns were retained for the new structure.

Under the Beeching recommendations this line had been closed to passenger traffic and retained only for freight working, so that it was possible to convert the existing double track to single, and this made possible the method of reconstruction chosen by the consulting engineers.

The new bridge consists of single-track 'half-through' welded steel-plate side girders with a deck built up of precast units consisting of steel cross-joists encased in concrete (fig 26).

The new girders rest on welded steel cill beams placed on top of the original cast-iron piers.

To enable these cill beams to be placed, first of all the new single track was laid at a higher level—466 mm—than the original, and the ends of the old girders modified by shortening the end posts and altering the end panel bracing to form a sort of 'step'. To enable this to be done, the girders were supported on temporary gantries attached to the piers which were high enough to permit traffic to continue. The new single track was supported on temporary way beams. After the cill beams had been inserted the old girders were lowered on to them, giving a further 330 mm extra working depth. The work of installing the new plate girders then proceeded, further temporary gantries being secured to the tops of the old girders to place the new sections. The new deck sections were placed by travelling gantry working along the flat tops of the new side girders. These gantries were also used to remove the old girders which were cut up into manageable lengths and loaded on to railway wagons. The old deck was also cut up and taken away by barges in the river.

A photograph of the operations is reproduced.

Each girder span was dealt with in turn so that the minimum amount of traffic interference was involved.

Rendel Palmer & Tritton were the consulting engineers to the then Chief Civil Engineer of the Eastern Region, Mr A K Terris, under whose overall supervision the work was carried out. The main contractors were the Butterley Engineering Co Ltd, who have very kindly supplied the photographs. Thos Fletcher & Co Ltd of Mansfield were the subcontractors.

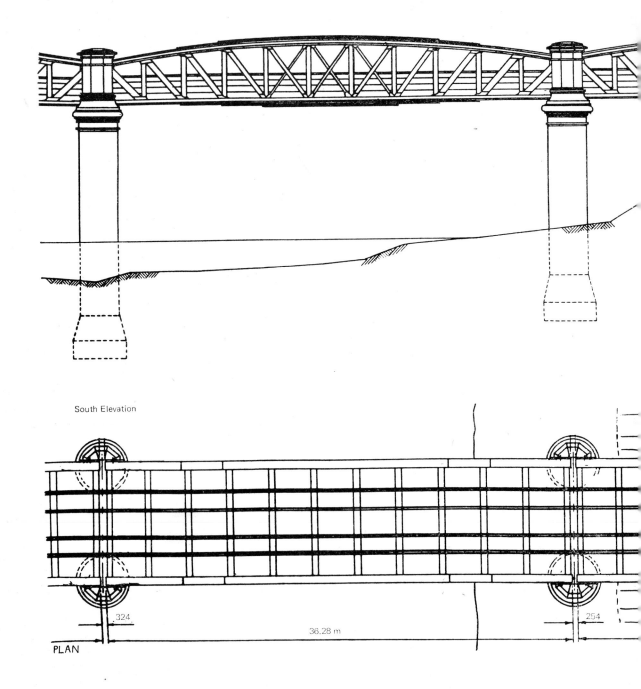

South Elevation

.324

36.28 m

254

PLAN

Fig 24   *Fledborough Viaduct: old bridge, B.R. Eastern Region*

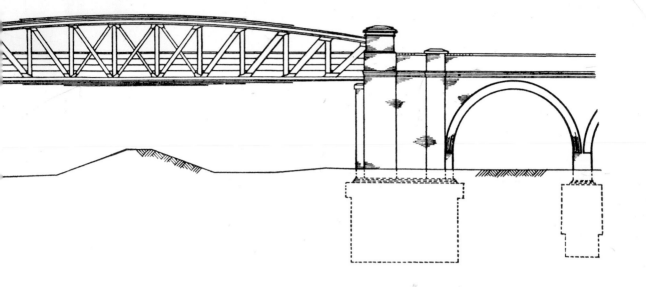

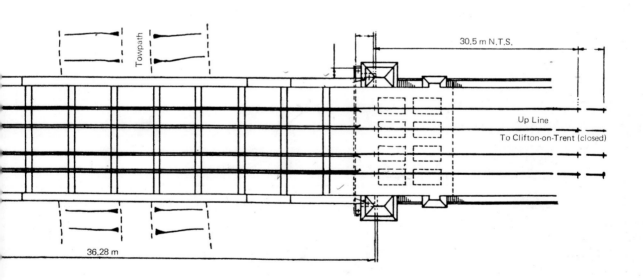

Towpath

30.5 m N.T.S.

Up Line

To Clifton-on-Trent (closed)

36,28 m

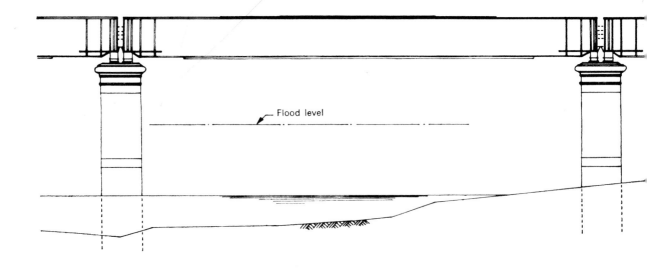

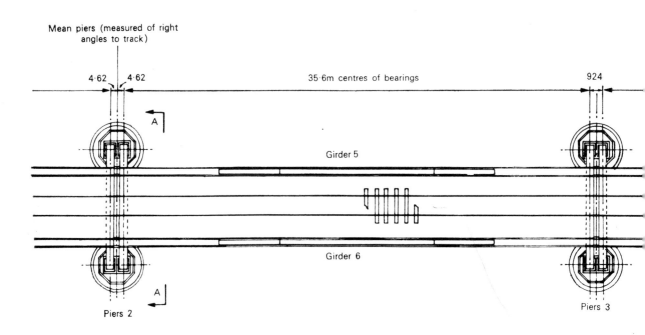

Fig 25 *Fledborough Viaduct : reconstruction, B.R. Eastern Region*

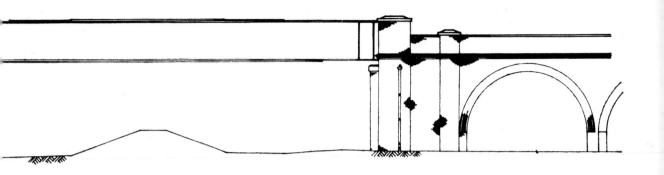

ELEVATION

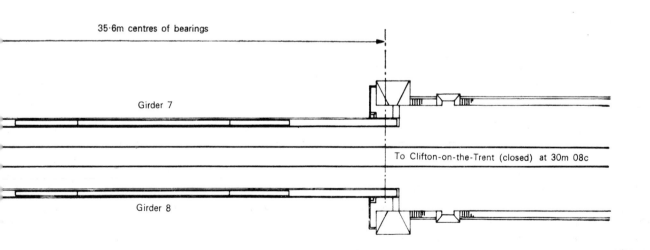

35·6m centres of bearings

Girder 7

To Clifton-on-the-Trent (closed) at 30m 08c

Girder 8

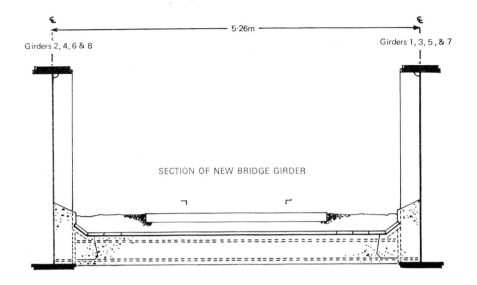

5·26m

Girders 2, 4, 6 & 8

Girders 1, 3, 5, & 7

SECTION OF NEW BRIDGE GIRDER

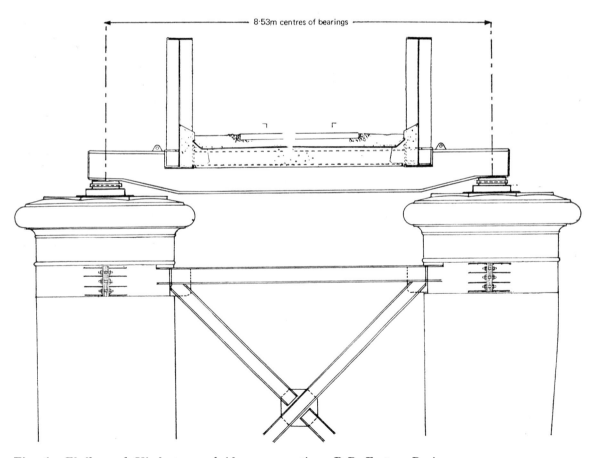

8·53m centres of bearings

Fig 26   *Fledborough Viaduct: new bridge, cross-sections, B.R. Eastern Region*

Plate 3.13   *Hyde Lane Bridge, B.R. Western Region*

Plate 3.14   *Fledborough Viaduct reconstruction, B.R. Eastern Region*

4 *Some notable concrete bridges in the UK*

In this chapter it is proposed to describe some of the interesting and sometimes unique bridges built on British Railways since the war, making use of prestressed concrete techniques. The factors which have assisted in the development of prestressed concrete in Britain have already been discussed in the Introduction. Some bridges are dealt with at greater length than others largely to bring out some of the early experiences with prestressed concrete and which helped to establish standards which have since produced some remarkable results, particularly on the London Midland Region of British Railways. Many of the new designs are also elegant and are aesthetically satisfying. There is no doubt that prestressed concrete lends itself to a certain slimness, and it is pleasing to note how advantage has been taken to produce designs which combine utility with good appearance.

The Adam Viaduct

This viaduct near Wigan on the LM Region was the first example of prestressed concrete construction making use of prestressed concrete precast beam units. It was a reconstruction of an original timber bridge in four spans of 9 m length. The abutments were masonry

Plate 4.1 *Adam Viaduct, B.R. London Midland Region*

Plate 4.2   *River Esk Viaduct, B.R. London Midland Region*

and the timber piers rested on masonry foundations.

The new piers are in reinforced concrete resting on the original foundations supporting the new deck consisting of prestressed concrete I section beams 81 cm deep, with a top flange 52 cm wide, a bottom flange 41 cm wide and a web thickness of 10 cm. There are eight lines of these beams placed side by side under the two tracks and side parapets. They are tied together transversely by 32 mm dia high tensile rods. The prestressing reinforcement consists of 5 mm dia wires of 15·5 tons/cm² ultimate strength. The beams were cast at the precast concrete depot at Newton Heath. The bridge was completed in 1947.

The River Esk Viaduct

Another early prestressed concrete bridge is that over the River Esk, shown in the accompanying photograph. It is located between Floriston and Gretna Junction, and was built in 1961 to partially replace the original steel structure, the piers of which were showing signs of deterioration due to the severe flooding to which the Esk Valley is subject.

It has thirty-seven spans of which those over the river itself are 10·2 m long, and the rest on either side over the flood area are 7·5 m long. The foundations for the piers consist of rows of twelve bored piles, 482 mm dia, which are capped by reinforced concrete beams.

The new bridge had to be wider than the old, some parts of which were retained, including masonry piers and some steelwork.

New construction consisted of inverted Tee precast prestressed concrete beams with concrete placed between to form a monolithic deck. Where old steel beams were retained, a new reinforced concrete well deck was built. The work was under the control of the then Chief Civil Engineer LM Region, Mr A N Butland. The piles were supplied by the Cementation Co Ltd and the concrete beams by Costain Construction Ltd.

Three prestressed concrete bridges at Manchester

Three early instances of the application of prestressed concrete techniques are to be found on the Manchester–Stockport main line at Stockport Road and Hyde Road, and on the former Manchester–Altringham line at Fairfield Street, and were occasioned by the LMR electrification from Euston to Manchester.

Both Stockport Road and Hyde Road bridges required to be strengthened to deal with the heavier traffic envisaged under electrification, and two factors entered into the decision to replace the existing steel and iron girders with concrete. First it was essential to keep the depth of the bridge the same, and secondly the steel delivery position existing at the time was such that the two bridges could not have been completed in the time required. Both bridges are of interest in that none of comparable size and strength had till then been constructed in prestressed concrete. They are described in

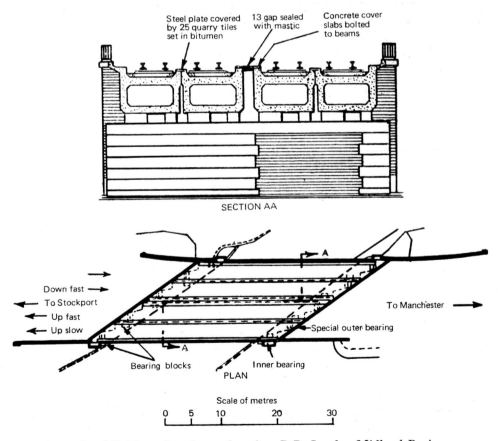

Fig 27 *Stockport Road Bridge : elevation and section, B.R. London Midland Region*

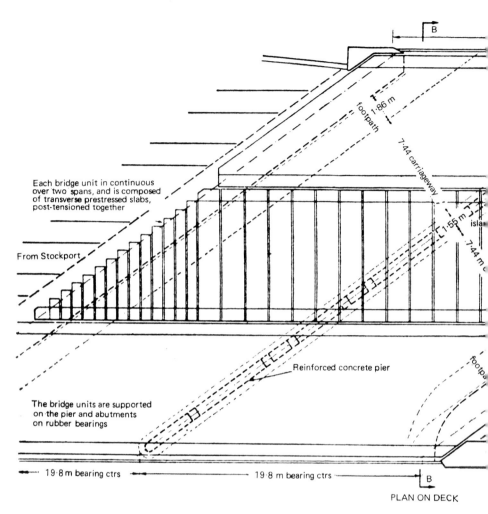

Each bridge unit in continuous over two spans, and is composed of transverse prestressed slabs, post-tensioned together

From Stockport

Reinforced concrete pier

The bridge units are supported on the pier and abutments on rubber bearings

footpath 1·86 m

7·44 carriageway

1·55 m

islan

7·44 m

footp

19·8 m bearing ctrs — 19·8 m bearing ctrs

B

PLAN ON DECK

Fig 28   Hyde Road Bridge : elevation and section, B.R. London Midland Region

Mr F Turton's paper to the ICE of September 1961, No 6357.

Stockport Road Bridge No 23

The original bridge on an acute skew was of somewhat 'hotch-potch' construction. The first part built in the early days of railways was a two-track structure in three spans with a central span of 17·6 m of cast-iron ribs resting on stone piers, and with two side spans which were 5·2 m stone arches. When the then LNWR main line was quadrupled in 1880 the second portion consisted of an outer steel box girder of 35 m span and an inner shallower steel girder. Both were continuous over three spans with a row of intermediate columns in line with the piers of the original bridge. Both rail and road traffic underneath were of high density, which made possessions difficult, and road possession could only be given at week-ends. It was decided to erect a temporary platform over the roadway on which to assemble the four concrete replacement girders one at a time, and to roll in each one successively from the Up side of the four-track route on temporary rolling-in trestles. Fig 27 shows the new bridge.

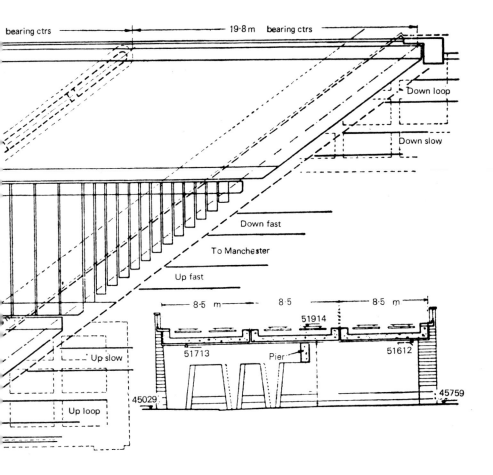

bearing ctrs ← 19·8 m → bearing ctrs

Down loop

Down slow

Down fast

To Manchester

Up fast

8·5 m   8·5   8·5 m

51914

51713   Pier   51612

Up slow

45029   45759

Up loop

The four concrete girders are of identical section except for the two outer ones, which incorporate a side parapet wall and a walkway. They were of hollow box section, 33 m between bearings, and with a total length of 43 m. Advantage was taken of the skew to reduce the mid-span bending moment. Each girder was cast in 35-ton sections 2·4 m long, which were brought by rail from the factory, transferred to road vehicles and taken to site where they were lifted by mobile crane on to the temporary erecting platform. Each girder consisted of sixteen 2·4 m sections, and these were placed end to end and the prestressing cables laid in grooves on the bottom and located in position by vertical diaphragms, and finally swept up to the outer ends of the girders. This operation occupied about five weeks. The sixteen pre-stressing cables were Magnel 72 strands of 19 mm dia steel wires.

The bridge bearings were rubber-steel sandwich type mounted on reinforced concrete pedestals. They were installed under traffic conditions ready to receive the girders. As there was not enough room on the erecting frame to accommodate all four girders at once, each one was rolled in separately, complete with deck and track laid and ballasted, commencing with the Up slow line. When the demolition of the old bridgework for the next track was completed, then this girder was moved over and replaced by a new one, and so on until all four were placed. Traffic was therefore maintained on at least two out of the four tracks.

Hyde Road Bridge No 30

This is a six-track bridge which crosses Hyde Road at an angle of 65°. The original bridge consisted of four steel box girders, two outer ones spanning the full width of 38 m, and two inner ones of shallower depth which rested on intermediate cast-iron pillars placed on the roadside kerbs. As the Manchester Corporation wished to create a dual-carriageway road, and to have the cast-iron pillars removed, a two-span continuous slab bridge with a central pier between the carriageways was adopted.

Actually the new bridge consists of three separate composite slabs each carrying two tracks, and resting on the two abutments and a central reinforced concrete pier which is built with six tapered legs. These slabs are of trough section 60 cm thick, with an upstanding ballast wall 77 cm wide and 59 cm high above the top surface of the slabs. They are 1·5 m wide except the end units, which are 0·7 m, as can be

Plate 4.3   *Fairfield Street Bridge, Manchester, B.R. London Midland Region*

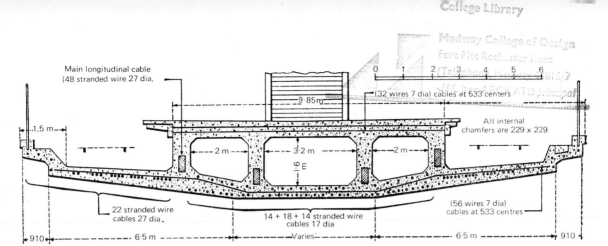

Fig 29   *Fairfield Street Bridge : section, B.R. London Midland Region*

seen in the plan view fig 28. They were precast in a factory and prestressed longitudinally with 13 mm dia stranded steel wires.

As at Stockport Road, an erection platform was built over the road on which the units were assembled and post-tensioned together with Magnel bars. The upstanding side walls were cast on and post-tensioned to form a strong double-track deck slab of some 600 tons weight. The three slabs are supported on rubber bearings.

After dismantling the old bridge and preparing the bearings, which was done under traffic conditions, the first slab was rolled, complete with two tracks, into the first position. When the next position had been prepared, and the next slab was ready, the first slab was moved over to the second pair of tracks, and the new slab rolled into the position previously occupied by the first. A similar procedure was carried out for the last slab. Three rolling-in paths were provided along each abutment and the central pier.

Fairfield Street Bridge

As part of the remodelling of the former London Road Station (now renamed Piccadilly) the two former platforms on the west side which served the electrified lines of the former Manchester South Junction and Altringham Railway were demolished, and a new island platform was built over Fairfield Street. At that time the MSJ & A was already electrified

on the 1,500 v dc system, and in order to avoid confliction with the new standard 25 kv ac system, which had been adopted for the Euston–Manchester electrification, it was decided to make the preceding station, Oxford Road, on the MSJ & A the terminus for electric trains to and from Cheshire. This decision, coupled with other extensive engineering works associated with the remodelling of London Road Station, made it possible to close the Oxford Road–London Road portion, thus making it possible to reconstruct Fairfield Street Bridge *in situ*, and so greatly simplify the work.

For this reconstruction the London Midland Region Civil Engineering Department conceived the idea of using the central island platform for the double duty of serving as a platform and bridge girder. This was done with the rail tracks carried on either side of the central main girder on cantilever construction.

The cross-section of the bridge is shown in fig 29 and indicates clearly the construction. It is a skew span of 51 m and has a width of 9·8 m. It was cast *in situ*, and the main beam was afterwards post-tensioned, and the number and sizes of the cables are shown. It was built between 1958 and 1961, and it is of particular interest that a design similar in principle was adopted for a much larger bridge at Besses o' th' Barn, also at Manchester, to be described later. The reasons which finally decided the civil engineers to use prestressed concrete were

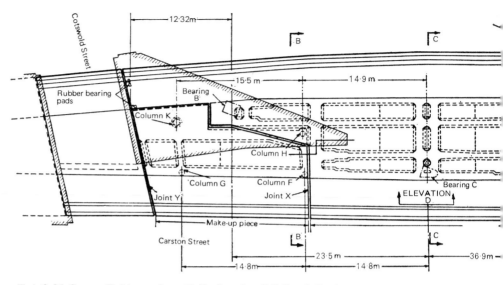

Fig 30 *Fairfield Street Bridge : plan, B.R. London Midland Region*

the earlier experiences with Stockport Road and Hyde Road bridges, where good use had been made of the skew to reduce the bending moments at the centre of the span, and where shallow depth was essential, and the delivery of steel-work unfavourable. Conditions were complicated by the very extreme angle of skew, and the fact that the structure was required to support a luggage lift, a combined pedestrian and luggage bridge, and a stairway. In fact, the well required for the lift was, in itself a further complication. The central main girder design had also been decided on after trying out various designs of through bridge which had been found to be either impracticable or undesirable, and it was decided to design the island platform as a hollow slab of full platform width, and with a depth equal to the distance from platform level to bridge soffit. This slab is supported at four points, and its torsional resistance is adequate to support the cantilevered tracks on either side. Two scale models, one to 1/48 scale made of Araldite, and the other to 1/24 scale in concrete, were tested at the Imperial College of Science and provided further design data.

The plan of the bridge, fig 30, shows it to consist of three parts, of which the longest spans Fairfield Street, the second Carston Street, and the third Cotswold Street. The main portion is carried on four reinforced concrete columns, two of which, A and B, carry 2,500 tons each, and the other two, C and D, 1,000 tons each. All the columns rest on concrete foundations which are laid on a bed of sand-stone at a depth of 7·5 m. Column A has a fixed base, but column B has a hinged base which permits longitudinal movement. The bearings at the top are 'Mehanite' castings with spherical surfaces. Columns C and D are hinged on rubber bearings. Fig 31 shows an elevation and section of column D. The portion of the bridge over Carston Street is supported on three columns and partly on triangular abutments on rubber bearings. The slab over Cotswold Street rests on rubber bearings on triangular abutments.

The total weight of the bridge is 3,000 tons, and the applied loading is 2,800 tons.

Fine-aggregate rapid-hardening cement was used with a mix of 1/1·53 by weight. The post-tensioning cables were Magnel bars 29 mm and 6 mm dia, finally stressed to 8 tons/cm².

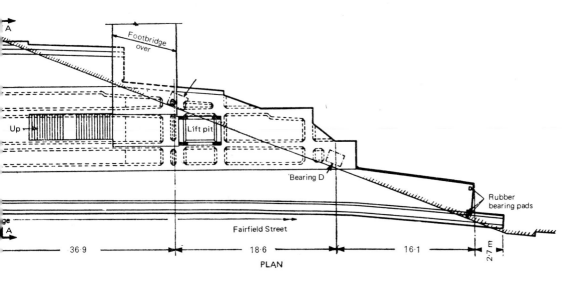

PLAN

Macalloy bars 29 mm dia were used for shear reinforcement in the webs and diaphragms.

The Penrith Bridges

These bridges are the first of a series of large prestressed concrete bridges of unusual design produced by the Bridge Section of the Civil Engineering Department of the London Midland Region. Actually the first notable design was Mossband Bridge near Carlisle. This was a road bridge crossing the railway and was described in Chapter 1. The three Penrith bridges were required for the construction of the Penrith bypass on the M56 motorway, are located about 4 km ($2\frac{1}{2}$ m) south of Penrith Station, and are known as Clifton, Skirsgill, and Penrith Station.

All are prestressed or post-tensioned designs, and were built in advance of the road by the side of the tracks, and subsequently moved into position by the then new sliding technique. Clifton and Penrith Station are hollow concrete box girders of similar design, 121 m and 111 m long, and are of trapezoidal cross-section post-tensioned. Skirsgill is a prestressed concrete slab 29 m long. During the design

period consideration was given to a steel beam of similar shape, like Trowell Bridge already described in Chapter 3, but final costing showed the concrete design to be the more economical. One of the advantages of the trapezoidal section is good torsional rigidity.

Fig 32 shows the elevation of one end of Clifton Bridge, a cross-section through one of the piers, and a plan of the abutment end.

Fig 33 shows a cross-section through Penrith Station bridge. This is really two bridges in one. The double-track bridge carries the two main-line tracks and is exactly similar to Clifton. The single-track bridge carries the Keswick branch which veers off to the west. Both bridges are entirely independent of each other. In all the bridges transverse vertical strengthening diaphragm walls are provided at intervals throughout the span. These also serve to maintain the post-tensioning cables in the correct alignment with respect to the longitudinal axis.

The bottom flange of Clifton is slightly cambered, and the depth is 3·4 m over the piers and 2·5 m at the centre of the span. In the case of Penrith, the depth is constant through-

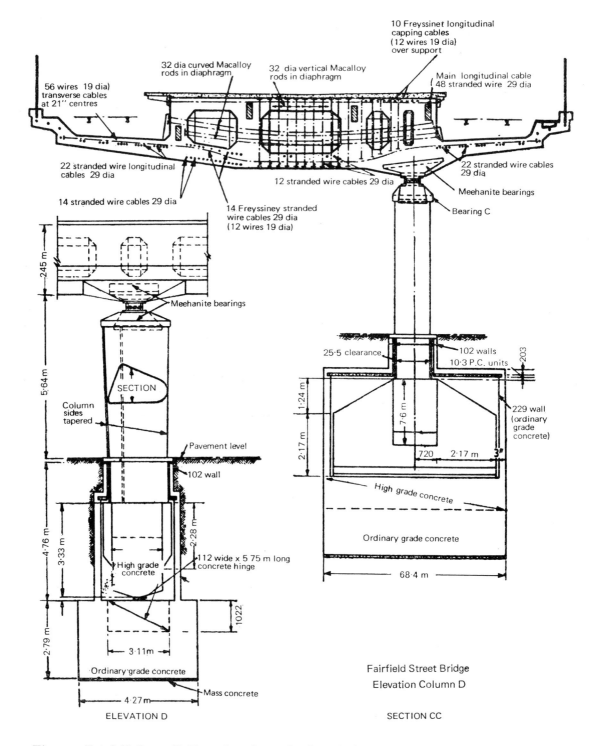

10 Freyssinet longitudinal capping cables (12 wires 19 dia) over support

32 dia curved Macalloy rods in diaphragm

32 dia vertical Macalloy rods in diaphragm

Main longitudinal cable (48 stranded wire 29 dia)

56 wires 19 dia) transverse cables at 21" centres

22 stranded wire longitudinal cables 29 dia

14 stranded wire cables 29 dia

14 Freyssiney stranded wire cables 29 dia (12 wires 19 dia)

12 stranded wire cables 29 dia

22 stranded wire cables 29 dia

Meehanite bearings

Bearing C

245 m

Meehanite bearings

5·64 m

SECTION

Column sides tapered

Pavement level

102 wall

4·76 m

3·33 m

2·28 m

High grade concrete

112 wide x 5 75 m long concrete hinge

1022

2·79 m

3·11m

Ordinary grade concrete

4·27 m

Mass concrete

ELEVATION D

25·5 clearance

102 walls

10·3 P.C. units

203

1·24 m

7·6 m

229 wall (ordinary grade concrete)

2·17 m

720

2·17 m

3"

High grade concrete

Ordinary grade concrete

68·4 m

Fairfield Street Bridge
Elevation Column D

SECTION CC

Fig 31　*Fairfield Street Bridge: elevation and column 'D', B.R. London Midland Region*

Plate 4.4 *Clifton Bridge, Penrith, B.R. London Midland Region*

out at 3·4 m. Both bridges consist of a central span with two 31 m approach spans. The central span of Clifton is 59 m and Penrith is 49 m. The double-track bridges are 9·4 m over parapet walls, and the single-track bridge is 7 m, which includes a footway. In the Clifton bridge there are thirty-six main post-tensioning cables, each consisting of 12/15 mm strands, and the total post tension was 6,800 tons. In order to avoid the possibility of friction at the diaphragms preventing the developing of the full tension, initial tension was applied at one ending of the girder, and the full tension at the other. This process was reversed for alternate cables. After stressing, the cables were encased in concrete as an anti-corrosion precaution. Suitably spaced manholes provided in the bottom of the girder enabled this to be done. Originally it had been proposed to build

the two long bridges from precast sections brought to site, but the contractors, Messrs Leonard Fairclough, decided to build them as complete monolithic units. The moving in of bridges of this size weighing some 2,500 tons was facilitated by the use of the sliding technique. Clifton Bridge was moved in in May 1967 during a 54-hour week-end possession of the two main lines.

The two long bridges are at the south end mounted on two fixed rota bearings and at the north end on four laminated rubber bearings which allow free longitudinal movement.

The elevation and cross-section of the other bridge, Skirsgill, is shown in fig 34. It is a good example of a modern prestressed concrete slab bridge. It is 29 m between abutment walls, 13 m wide, and 1·67 m deep. It carries three tracks, and, as the cross-section is on a curve

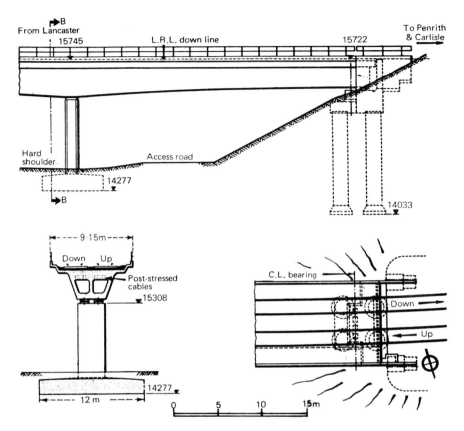

Fig 32   *Clifton Bridge : elevation and section, B.R. London Midland Region*

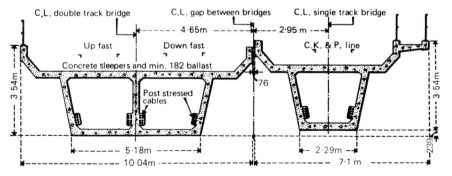

Fig 33   *Penrith Station Bridge : section, B.R. London Midland Region*

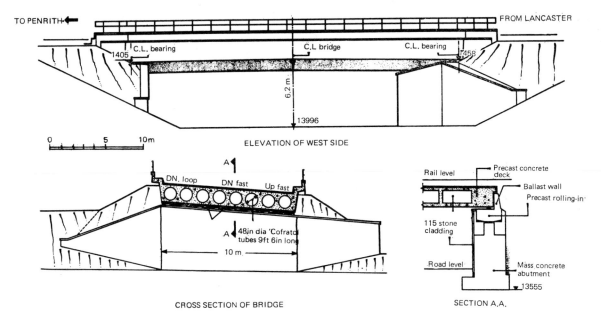

TO PENRITH◄   FROM LANCASTER

C.L. bearing   1405   C.L bridge   C.L. bearing   1458

6.2 m

13996

0   5   10m

ELEVATION OF WEST SIDE

A

DN. loop   DN fast   Up fast

A   48 in dia 'Cofrato' tubes 9ft 6in long

10 m.

CROSS SECTION OF BRIDGE

Rail level   Precast concrete deck

Ballast wall

Precast rolling-in

115 stone cladding

Road level

Mass concrete abutment

13555

SECTION A.A.

Fig 34   *Skirsgill Bridge : section, B.R. London Midland Region*

with pronounced super-elevation, special precast concrete rolling-in beams were built which took account of the super-elevation.

The bankseats and piers for these bridges were constructed by the sheet pile wall method, in which lines of steel sheet piles are driven transversely across the tracks at the appropriate distance apart, and bridged with beams to support the tracks between the sheet pile walls so formed. Excavation work then proceeded under normal traffic conditions and the abutment walls and pier foundations constructed. The two piers for Clifton and Penrith are 2·97 m × 1·4 m × 8 m high. They rest on reinforced concrete slabs 12 m × 7 m × 1·9 m thick. Concrete moving-in walls were built opposite each abutment and pier on which the bridges were erected and subsequently moved in. The actual moving of these bridges has already been described in Chapter 2. A number of photographs of the construction work are shown.

Messrs Leonard Fairclough were the contractors for the bridge superstructures, and the substructure work was done by the British Rail Direct Labour Organisation. The whole work

was under the overall supervision of Mr W F Beatty, Chief Civil Engineer of the London Midland Region.

Besses o' th' Barn Bridge: Manchester–Bury line

This bridge, shown in fig 35, is a remarkable design forming part of a three-tier bridge arrangement where the Lancashire–Yorkshire motorway crosses under the Manchester–Bury electrified line, and also under the Bury Old Road. The railway is at the top, and the motorway at the lowest level. New road and rail bridges were required, the road bridge being built by the Lancashire County Council.

In this area subsidence due to coal-mining is a major problem, and over the next twenty-five years about 3·4 m subsidence is expected. To cope with this it was decided to adopt a three-point support system combined with foundations capable of individual adjustment in height. This led on to the design form of a central reinforced post-tensioned concrete 'spine' with the two tracks carried on decks cantilevered out from it on either side. This enabled the

Plate 4.5   *Clifton Bridge under construction, B.R. London Midland Region*

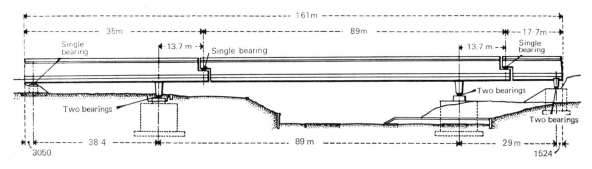

Fig 35   *Besses o' th' Barn Bridge : elevation, B.R. London Midland Region*

Plate 4.6 *Clifton Bridge after sliding-in operation, B.R. London Midland Region*

Plate 4.7 *Besses o' th' Barn Bridge, Manchester, B.R. London Midland Region*

Plate 4.8 *Besses o' th' Barn Bridge: end view, B.R. London Midland Region*

spacing of foundations for the two bearing end of the spine to be kept within a desirable limit, the other end being on a single support, all three being adjustable in height if required.

The resulting design is reminiscent of Fairfield Street Bridge at Manchester. It consists of three spans of 41 m, 91·5 m, and 31 m length. Each span has a single bearing at one end, and a two-leg trestle at the other, the trestle being cast integrally with the spine. The two independent foundations for the trestle legs can be adjusted both vertically and longitudinally. As can be seen from fig 36, the single bearing end of each span rests on a 14 m extension beyond the trestle leg on the span in front.

Thus the short 17 m girder rests on two bearings on the north abutment, and on a single bearing on the extended end of the long 91·5 m central girder. This in turn rests on its two trestle leg bearings at the north end, and on a single bearing on the end of the 56 m southern girder. This rests on two trestle bearings and on a single bearing on the southern abutment.

The cross-section of the spine is the same for all three girders. It is a massive hollow beam 6·8 m high and 4 m wide at the top and 0·91 m thick. At deck level it is 3·8 m wide with 23 mm thick side webs sloping inwards to meet the top flange. Details are shown in fig 36. It is built up of a number of precast units brought to

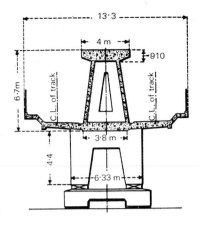

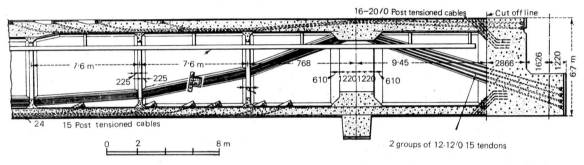

Fig 36 *Besses o' th' Barn Bridge : sections, B.R. London Midland Region*

site and erected in position on trestling. The longitudinal post-stressing tendons comprise twenty-four cables each consisting of twelve 15 mm strands. Vertical diaphragm walls are provided at intervals of 7·62 m which serve to maintain the correct profile of the longitudinal cables. Similar cables are provided in the top and bottom flanges of the spine, with sixteen in the top and twenty-four in the bottom. An internal runway beam is provided along the interior for servicing and maintenance, and was also used to encase the cables in concrete.

The two deck slabs were cast on to the spine and were prestressed transversely. The width over the two decks is 13·7 m. The composite construction of the spine and decks was designed to take account of the torsional stresses when unsymmetrical load is imposed by one train. In order to check the stresses developed in the

stepped end blocks of the overhanging ends, a 1/8 scale model was tested at the British Rail Research Department at Derby.

The bridge was built along its final alignment, and to enable this to be done the two existing tracks were converted to a single track which was diverted over a temporary bridge. The long central span was erected in its final position, while the southern span, although built on its final alignment, was erected at sufficient distance away to enable the stressing of the cables in the central span to be carried out. The southern abutment and pier were built on special foundations which permitted this span to be moved on special sledges to join up with the central span. The much shorter northern span was constructed in ordinary reinforced concrete in its final position. The completed bridge weighs some 7,000

Plate 4.9 *Besses o' th' Barn Bridge: assembly of main 'spine', B.R. London Midland Region*

tons. The contractor was Leonard Fairclough, and the work was under the overall supervision of Mr W F Beatty, the Chief Civil Engineer of the London Midland Region British Railways.

Five photographs of this remarkable bridge are reproduced.

The Frodsham Bridges

Two of the three bridges at Frodsham on the LM Region are interesting not only on account of their original design conception but also on account of the method of erection. The two are Nos 4A and 5, the first is on the Chester to Runcorn Branch, and No 5 on the Crewe to Runcorn portion of the LMR main line to

Liverpool. The design has two prestressed reinforced concrete Bow String arch ribs with a suspended deck. Fig 37 shows elevations and a plan of the two bridges which are identical. The two arch ribs are inclined inwards and joined at the centre and the prestressed concrete deck is suspended from them by inclined steel cables. The main span which rests on reinforced concrete columns is 64.2 m long with two approach spans of 14.2 m. The centre span is cantilevered out 3.1 m at each end and is stepped to make a bearing for the two approach spans. These are of identical cross-section to that of the main deck. They are 11 m long deck troughs, and at their outer ends they rest on two bored piled 1.8 m dia. The mean height

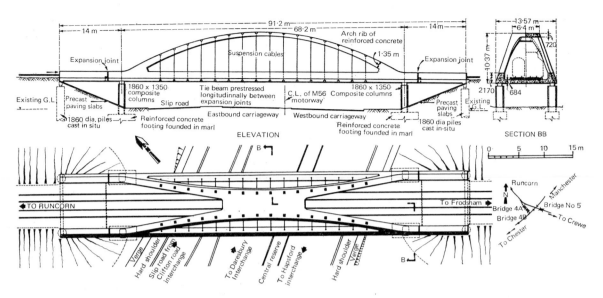

Fig 37   *Frodsham Bridges No 4A and 5: plan and elevation, B.R. London Midland Region*

Plate 4.10   *Frodsham Bridge No 4A, B.R. London Midland Region*

Plate 4.11　*Frodsham Bridge No 4A : end view, B.R. London Midland Region*

of the arch is 10·8 m, giving a height:span ratio of 6:1. The design was undertaken jointly by Messrs Husband and British Railways. The total width of the deck is 13·8 m.

All the columns and piles were constructed clear of the railway so that no interruption of traffic occurred until the bridge, which was erected at the side of the tracks, was ready to be moved in. Fig 38 shows how the bridge was built. In addition to the two columns on either side of the railway which form the foundations for the main girder, a temporary bored pile and trestle was erected opposite each of the permanent columns. On top of these were placed the combined moving-in and bridge-erection beams. The four stages in the construction are clearly indicated in the figure. The moving in of the bridge, which was by sliding on special sledges, occupied a single week-end.

The third bridge, No 4B, located near No 4A, is a prestressed concrete slab, and is shown in fig 39. It is 39 m long and 8·5 m wide. It is supported on eight bored piles 1·8 m dia, all sunk outside the track, as was the case for the other two bridges. No interruption to traffic was involved during the construction until the bridge was moved into position by sliding during a week-end possession.

These three bridges provide an example of

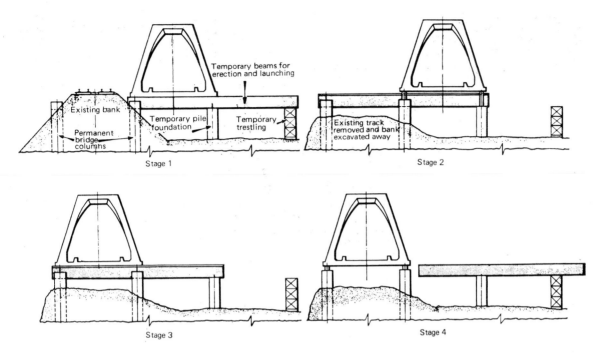

Fig 38 *Frodsham Bridges No A4 and 5 : method of erection, B.R. London Midland Region*

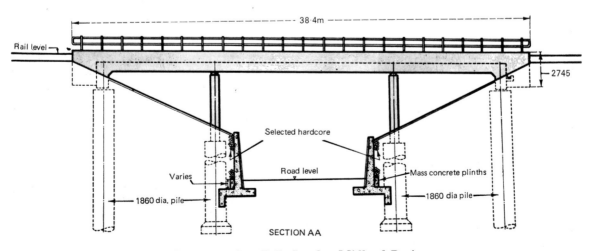

Fig 39 *Frodsham Bridges No 4B : section, B.R. London Midland Region*

Plate 4.12　*Canal Bridge No 11, Leeds City Station, B.R. London Midland Region*

how the minimum of traffic interference can be achieved when conditions are favourable.

Canal Bridge at Leeds

The rebuilding and the concentration of all railway lines in and out of Leeds into the new City Station at Leeds required the reconstruction of two separate bridges over the Leeds–Liverpool Canal as a single bridge.

It carries four tracks, and is a skew two-span bridge, and is a typical modern prestressed precast concrete structure. The beams are of I section prestressed with pre-tensioned wires. and they vary in length from 15 m to 21 m, and in depth from 1·03 m to 1·33 m.

The parapets are built up from concrete blocks, as can be seen in the illustration which shows how precast concrete troughs are incorporated in the parapets and walkways.

The bridge was constructed in two stages, each successively replacing one of the two earlier bridges. It was designed in 1960 by the Chief Civil Engineers of the North Eastern (now the Eastern Region). The contractors for one stage were Yorkshire Hennebique Ltd and for the second stage Wellerman Bros Ltd. The precast concrete beams were supplied by Anglian Products Ltd.

Bridge No 3A at Wandsworth

This new bridge on the Southern Region was built to allow a new GLC road to be constructed under the SR mainline, which is on an embankment 6·2 m high. It carries four main-line tracks and three sidings and is the largest example to date of the use of 'pipe-jacking' for the construction of the abutments and pier. It is in two spans 15 m long and has a total

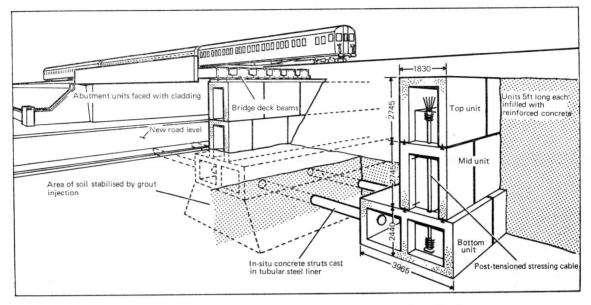

Fig 40 *Bridge No 3A at Wandsworth: isometric drawing, B.R. Southern Region*

Labels in figure:
Abutment units faced with cladding
Bridge deck beams
New road level
Area of soil stabilised by grout injection
In-situ concrete struts cast in tubular steel liner
1830
2745
Top unit
Units 5ft long each infilled with reinforced concrete
2745
Mid unit
2440
3965
Bottom unit
Post-tensioned stressing cable

width of 29·6 m. Each span comprises thirty-three prestressed concrete beams, and two precast concrete parapet beams. On one side of the main structure there is a footbridge also constructed of precast units. The two abutments and the central pier are built up of hollow rectangular section reinforced concrete segments 'pipe-jacked' through the embankment under the tracks under traffic conditions. As at Haresfield Bridge on the Western Region, the abutments and the pier consist of three tiers of segments which are driven as shown in fig 40. The lowest row is formed from segments 4 m × 2·5 m × 25 cm thick laid on their sides. The next two rows consist of segments 2·8 m × 1·9 m in section placed upright, in the case of the abutments on the outer edges of the lowest tier, and in the case of the pier in the centre. This creates an L-section abutment, and an inverted Tee-section for the pier. The larger bottom units have a central wall dividing it into two compartments. All the units are 1·5 m long. The units were tied together internally by post-tensioned stressing cables, and filled with mass concrete. After driving it was found necessary to stabilise the ground by chemical

injection, particularly under the pier. In addition, to cope with lateral forces on the abutments, they were tied together by concrete-filled steel tubes 0·9 m dia under the road. The bridge beams are 78 cm wide and 98 cm deep at the outer ends. They are of a broad inverted Tee-section. The deck is completed by filling the spaces between the beams with concrete.

As there were seven tracks, it was possible to lay the beams for one track by crane while traffic continued on the others. The beams are simply supported over the central pier, and at the abutments they are secured by steel dowel pins let into the bearing caps, and projecting into preformed holes in the undersides of the beams themselves, and into which they were grouted. The same was done at the pier, except that the pins were not grouted, so as to permit a measure of freedom of movement.

The beams were standard BR precast prestressed concrete type, and after they had been placed they were covered with a damp-proof membrane and stressed together transversely.

Mr R E Evans is the Chief Civil Engineer for the Southern Region under whose supervision

Plate 4.13   *Rugby Fly-over, B.R. London Midland Region*

the work was carried out. Tube Headings Ltd did the 'pipe-jacking' of the foundations.

The Rugby Fly-over

This major work was built in 1962 using precast prestressed concrete beams supported on prefabricated concrete trestles. The photograph shows the construction. It enables Up trains from Birmingham to cross over the north-going main lines to Crewe, and eliminates the previous confliction of traffic on the old flat junction. Although planned before electrification, it contributes materially to the acceleration of the London to Birmingham services

which electrification has made possible. When electrification construction commenced, the overhead equipment standards were added, and some additional clearances had to be provided in 1963 at the point where the photograph was taken.

The Bletchley Fly-over

A photograph is included of another fly-over constructed at Bletchley on the LM Region which enables trains from the line from Cambridge to pass over the main line to the north without traffic conflict. The picture shows it under construction. It is built of precast prestressed concrete beams laid on reinforced concrete columns and cross-beams.

Pilgrim Street Bridge, Newcastle upon Tyne

This important bridge on the Eastern Region carries the four track East Coast route to Scotland over Pilgrim Street into Newcastle. It required to be widened to allow for a new inner ring road, and some interesting erection problems had to be solved to enable all four lines to be kept open to traffic.

On this section of the route the tracks are carried on a brick arch viaduct on their way to the River Tyne Bridge. The existing bridge over Pilgrim Street was a reinforced concrete structure with granite-facing blocks, and steel girders within the structure. This bridge had to be demolished and replaced by a longer structure which consists basically of four post-tensioned, precast concrete box beams,

one for each track. The road and rail traffic is dense, which added to the construction problems.

The whole scheme was put into the hands of consulting engineers Messrs W S Atkins & Partners, and work commenced in 1966. It was obvious that in order to keep all four tracks open a temporary diversion would be necessary, and it was decided to build a two-track temporary trestle bridge on the north side. This was used for the Up and Down main lines, and the Up and Down Tynemouth lines were slewed over into the position previously occupied by the main lines. This enabled demolition work to be started on the two southern most tracks.

The trestle bridge was some 232 m long, built of steel, which proved to be cheaper than timber. As it was required for a minimum period of two years, drilling was kept to a

Plate 4.14   *Bletchley Fly-over, B.R. London Midland Region*

minimum so as to increase the recovery value of the steel, and where possible Lindapter clips were used. The rails were carried on deck timbers with rubber mats under the chairs.

At the connections to the existing viaduct at the two ends special precautions were taken to deal with the transverse forces due to the track curvature, and the track decks were tied back by long steel bolts to mass concrete anchorages cast against the existing arches. The stages of the reconstruction are shown in fig 41. Once the tracks on the trestle bridge were in operation, and the Tynemouth lines slewed over, the demolition of the two southernmost tracks was begun after extensive shoring work to contain the unopposed arch thrusts, and

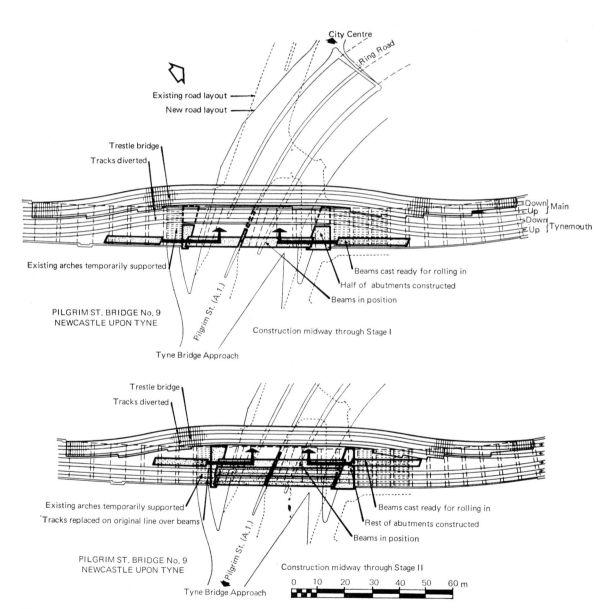

Fig 41　*Pilgrim Street Bridge : method of construction, B.R. Eastern Region*

Plate 4.15 *Pilgrim Street Bridge, Newcastle upon Tyne, B.R. Eastern Region*

the free edges of exposed arches. At first the demolition was carried out with jack hammers, but use was subsequently made of controlled explosives.

The half-abutments were built and the central pier in Pilgrim Street, and in the space created by the track slewing the half-girders were cast. Finally each half-girder after being post-tensioned was launched out towards the central pier. The leading ends of the girders were supported during the launching by a moving trestle located in the street below until they finally reached the pier, and then jacked down on to the bearings. When two girders had been thus placed the two southern tracks were laid

and placed in service, and this was the end of stage one.

The same procedure was employed for stage two, and on completion the temporary trestle was removed, and thus the whole reconstruction was completed without interruption to traffic except for some line possessions at the outset when the temporary decks were being tied back to the viaduct (fig 41).

The abutments for the new bridge are of cellular reinforced concrete on new foundations supported on piles driven down to firm boulder clay 9 m below ground level. They were tied in to the existing arch piers. The box section post-tensioned girders are 1·7 m deep, and 23

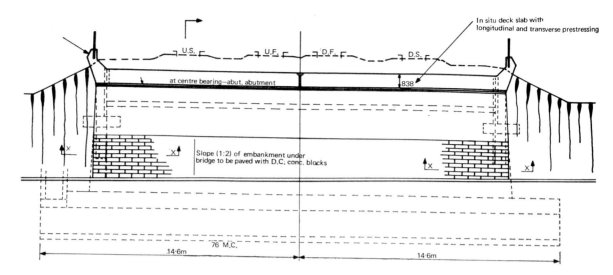

Sectional elevation D–D (south abutment).

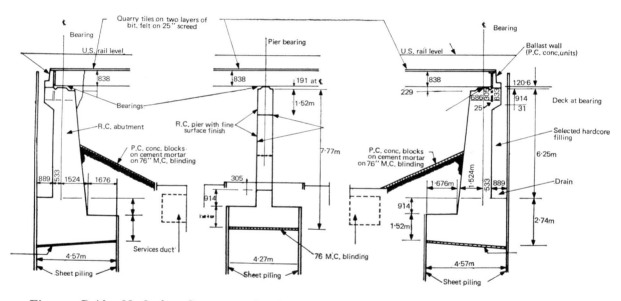

Fig 42   *Bridge No 85A at Stevenage: details, B.R. Eastern Region*

m and 28 m long, and rest on a three-bearing system with two bearings on the pier, and a single sliding bearing on the abutments. Although each girder is designed as an independent unit, *in situ*, shear keys are incorporated which simplify water-proofing and provide additional safety. The bridge was completed in 1968.

Bridge No 85A at Stevenage

This prestressed concrete slab bridge was completed in 1969 and is located about 0·6 km south of Stevenage Station on the Eastern Region main line from King's Cross. It was built on behalf of the Stevenage Development Corporation to allow for the construction of the Fairlands extension road which forms part of the Stevenage Main Road Development Scheme. The route here is four track on a low embankment some 2 m above ground level, and the new road is in a cutting passing underneath. The fact of there being four tracks enabled traffic to be maintained on two tracks during construction work.

It has an overall width of 25·2 m, and an overall length of 55·3 m, and consists of two clear spans of 20 m with a central pier.

Each pair of tracks is carried on a separate prestressed concrete slab which is continuous over the central pier. This pier has four legs founded on a concrete base slab. The slabs were cast on either side of the tracks and were separately rolled in when completed. The parapet walls were cast on to the decks before rolling in, and the wall section is continued behind each abutment as a wing wall section for a distance of 7.6 m

The abutments and the pier are founded on stiff clay at a depth of 8·4 m. They were built on the sheet pile wall method with two such walls for each abutment and the pier. Driving of the piles was confined to one track at a time, leaving two tracks available for traffic at all times. The lines of piles were continued on

Plate 4.16 *Stevenage Bridge No. 85A, B.R. Eastern Region*

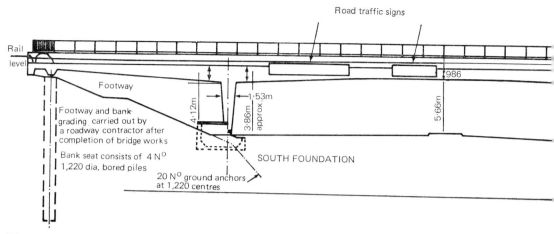

Rail
level

Footway

Footway and bank
grading carried out by
a roadway contractor after
completion of bridge works

Bank seat consists of 4 N⁰
1,220 dia. bored piles

20 N⁰ ground anchors
at 1,220 centres

4·12m

3·86m approx.

1·53m

SOUTH FOUNDATION

·986

5·66m

Fig 43   *Bridge No 1E at Derby: diagram and sections, B.R. Eastern Region*

Plate 4.17   *Bridge No 1E at Derby: under construction, B.R. London Midland Region*

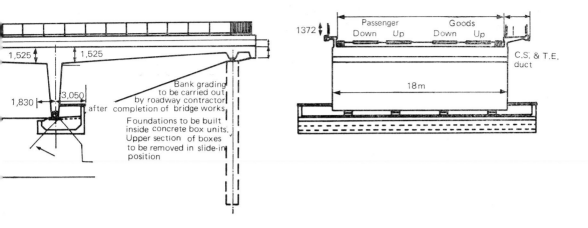

1,525   1,525

1,830   3,050

Bank grading
to be carried out
by roadway contractor
after completion of bridge works.
Foundations to be built
inside concrete box units.
Upper section of boxes
to be removed in slide-in
position

1372

Passenger   Goods
Down   Up   Down   Up

C.S. & T.E.
duct

18 m

Plate 4.18   *Bridge No 1E at Derby: prior to sliding-in, B.R. London Midland Region*

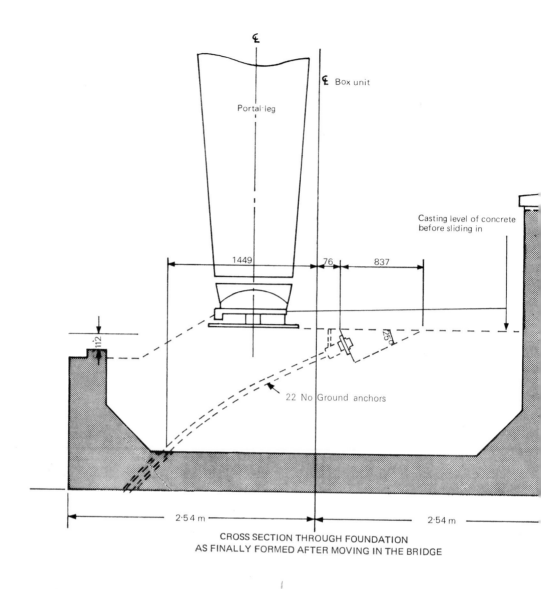

Fig 44   *Bridge No 1E at Derby : detail of foundation, B.R. Eastern Region*

either side of the railway to form support for the rolling-in beams on which each slab was built and subsequently moved into position. A total of 150 piles 15 m long, each weighing some 2½ tons, were driven. Each abutment is 8·8 m high with a base slab 30·4 m by 4·6 m and 2·8 m thick at the back, and 1·6 m at the front.

The pier is 8·6 m high, with a base slab 25·4 m by 4·3 m and 1·2 m thick (fig 42). The two prestressed concrete deck slabs are each 41·3 m long, 85 cm thick, and 12 m wide. They were prestressed longitudinally and transversely. At each abutment and on the pier they rest on twenty-six Andre steel reinforced rubber bearing

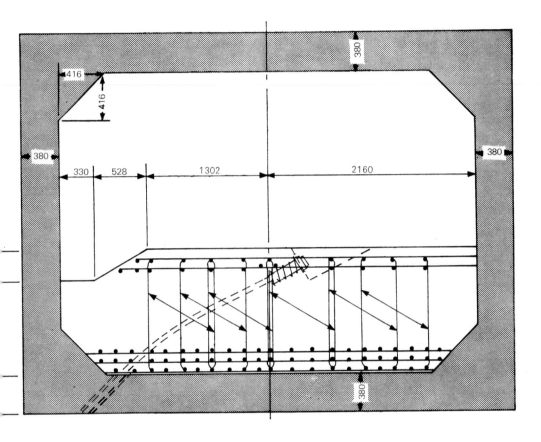

DETAILS OF GROUND BEAM
AS DRIVEN WITH LEG SEAT·FORMED WITHIN

Scale

blocks. On the pier these blocks are dowelled into the bearing seating and also into the deck by steel dowel pins, so that the deck is fixed in position on the pier and free to slide longitudinally on the abutments.

The deck carrying the two Down lines was moved in during the week-end June 7/9, 1969, and the slab carrying the two up tracks two weeks later. A speed restriction was imposed during the moving-in operation, and from December 1968 to April 1969 a limit of 90 km/h on the fast lines was in force, so that the traffic restrictions cannot be said to have been excessive.

Plate 4.19   *Bridge No 1E at Derby: completed, B.R. London Midland Region*

Leonard Fairclough Ltd was the contractor for the civil engineering work, and the overall supervision was under the Chief Civil Engineer of the Eastern Region, Mr A. W. McMurdo. This is an elegant modern prestressed concrete slab design, as the photographs show.

Bridge No 1E at Derby

This fine new design in post-tensioned concrete now, at the time of writing (1970), in course of erection is remarkable for several reasons. It is a monolithic reinforced post-stressed concrete portal structure with propped cantilever extensions which is unique, and has already attracted the attention of civil engineers abroad. Then, as the two legs are virtually hinged at the base

since they rest on spherical 'Glacier' bearings, the foundations are relatively small because they only receive direct horizontal and vertical reactions from the bridge itself, and as there is little pressure from the earth banks the overturning forces on the foundations are relatively secondary

A diagram of the bridge is shown in fig 43. It was designed by the Bridge Section of the LM Region's Chief Civil Engineers Department in conjunction with the Derby Corporation to enable a new ring road to pass under the railway, and, although not the least costly of several other designs which were considered, it was selected partly for aesthetic reasons as being in comformity with the environmental characteristics of the neighbourhood, and partly because a single clear span was desired across the new road.

The ground contours and the two-hinge principle of the structure led to an interesting application of the 'pipe-jacking' method of constructing the foundations.

These consist of a single row of hollow reinforced concrete units thrust through the railway embankment with the top sides made so that they could be removed and turned into open troughs. In this case, it was less costly than the two- or three-tier 'pipe-jacked' abutment foundations required for such bridges as Haresfield on the Western Region, or Wandsworth on the Southern Region already described. The hollow rectangular section units were 4·96 m wide, 3·7 m high, and each unit was 1·24 m long. The walls and floor were 38 cm thick, and the top 46 cm thick. The top was made so that it could be subsequently removed. There were twenty 1·24 m units in each drive, for which ten 100-ton jacks were used. For such a long drive (25 m) an intermediate jacking point was considered desirable, and nine additional 150-ton jacks were inserted at halfway, and the drive continued with the two sets of jacks operating together. The 'pipe-jacking' was done by Tube Headings Ltd. Before commencing the drive, some chemical consolidation of the ground was carried out by Soil Mechanics Ltd, and work commenced in April 1970.

When the two drives were completed, the work of constructing the bridge seatings went on inside the hollow beams, and at one end of the hollow 'tunnel' temporary concrete beams were laid on which the bridge was erected, and which later formed the sliding-in beams. The erection procedure was therefore as follows: When the bridge had been built the tops of the two foundation beams were removed, together with the top soil, and the whole bridge moved into position on sledges and jacked down on to the already prepared seatings constructed inside the foundation beams.

The cantilever ends of the bridge rest on two bored reinforced concrete piles sunk on either side and clear of the tracks, there was thus no interruption of traffic until the time came to slide the bridge in. When the bridge was in position on the leg foundations the cantilever ends were jacked upwards so that the bearings on the pile heads could be inserted, and the cantilevers lowered on to them. This operation had the effect of inducing an inward thrust on the leg bases to enable the necessary portal action to be regained.

The driving pits to house the jacking gear, constructed of sheet piling opposite each foundation beam, were 4·6 m square and 3·7 m deep, and suitable access for the installation of the jacks, and to enable the concrete units to be brought in, had to be created.

Fig 44 also shows a section through the bridge foundation, and it will be noted that twenty-two ground anchors were installed to resist the horizontal thrust of 500 tons acting at the leg bearings. The total length of the bridge is 77·7 m and the span between the leg bearings is 45·6 m. The cantilever spans at each end are 21 m long. The bridge carries four tracks, and is 17·4 m wide. On one side there is a 2·68 m wide footpath, and on the other a walkway 1·32 m wide, so that the overall width is 21·4 m. The clear headroom above the road is 5·6 m. The deck varies in thickness from 1·5 m over the legs to 1·08 m at the centre, and 62 cm at the ends of the cantilevers. The arrangement of the reinforcement and post-tensioning tendons is shown in fig 44. There are ninety-eight tendons each composed of eleven No 18 mm dia high strength strands which were tensioned up to 294 tons. Half of these tendons extend from end to end of the deck. The other half (forty-nine) are disposed over the portal portion in the lower part of the deck. Transverse and other reinforcement is composed of 25 mm and 19 mm bars.

The legs are tapered in side elevation being 1·55 m wide at the top, and 77 cm at the base. They are 5 m high, and the transverse length is 18·3 m. They are post-tensioned vertically by groups of four Macalloy bars 32 mm dia arranged in two rows, and the groups are spaced at intervals 36 cm apart throughout the width of the legs. The four Glacier

Plate 4.20 *Pipe bridge at Runcorn, B.R. London Midland Region*

spherical bearings for each leg are spaced 5 m apart.

The augur-driven piles on which the cantilever ends rest are 1·24 m dia and are located 18·6 m from the centre lines of the legs. There is thus an overhang of 2·18 m beyond the pile centre lines. A transverse reinforced concrete beam 62 cm deep is formed in the underside of the bridge deck over the piles, and this rests on sliding bearings on the pile heads.

The main contractor for the work was A Monk & Co Ltd, and the overall supervision of this interesting construction was under the Chief Civil Engineer for the London Midland Region, Mr W F Beatty.

Pipe bridge at Runcorn

The semicircular reinforced concrete arch built in 1961 at Runcorn to carry a water supply over the railway is an interesting small construction and is illustrated in the accompanying photograph. It comprises two three-hinge arches each constructed in two halves. The span is 10·3 m and the height above rail is 5·8 m. It is placed 3 m in front of the existing arch overbridge at this spot. It carries a 331 mm dia water main, and two 152 mm dia pipes.

Plate 4.21 *Roofing over Edge Hill cutting, B.R. London Midland Region*

Each arch segment is 610 mm deep by 305 mm wide, and the two half-arches rest on a combined foundation on each side of the tracks, and are connected at the centre by a hinged joint. They are separated laterally at 2·9 m intervals by *in situ* cast tie beams which rest on precast projections formed on the inner sides of the arch ribs. The bridge deck is formed by precast concrete units 152 mm deep, and 305 mm wide placed between the tie beams, and the whole is covered by a concrete mat 38 mm thick.

The parapet walls are 1·38 m high and are constructed of 100 mm thick precast concrete

units curved to the arch profile and bolted to the main arch ribs.

This bridge was designed by Mr G Penman, AMICE, Engineer to the Runcorn District Water Board, and cost £3,056.

Roofing over cuttings

Prestressed prefabricated concrete beams with their corrosion resisting properties, and relatively shallow depth are ideal for bridging over existing cuttings to form a base for building development above the railway. Photographs are reproduced showing a long 46 m concrete beam being laid over the cutting on the line out of Lime Street Station, Liverpool, over a former vent hole no longer required under electric traction. Another shows a 61 m length of cutting at Walsall being covered by 15-ton 21 m long prestressed concrete beams to provide the base for a building project. There were forty-nine of these beams, and they were not laid side by side, but spaced some distance apart and bridged in between by prestressed concrete planks, and the whole completed by a concrete slab 38 mm thick.

The rebuilt New Street Station, Birmingham, has a roof over the tracks constructed of pre-stressed precast concrete beams, on which a concourse, a shopping centre, and other amenities were built. Two photographs are included of the interesting reconstruction of Hampstead Road bridge just outside Euston Station occasioned by the remodelling of the tracks to suit the station reconstruction. An additional span had to be provided, and the work was carried out in two halves to enable road traffic to be maintained. The existing steel bridge was replaced by precast prestressed concrete beams resting on new reinforced concrete piers built *in situ*.

Plate 4.22   *New roof beams for New Street Station, Birmingham, B.R. London Midland Region*

Plate 4.23   *Reconstruction of Hampstead Road bridge during Euston Station remodelling, B.R. London Midland Region*

Plate 4.24 *Bridge No 3A, Wandsworth, B.R. Southern Region*

5 *Footbridges on British Railways*

From an aesthetic point of view railway footbridges with their high stairways, and comparatively short spans are not easy to design with elegance. In comparison with footbridges over roads, it is hard to find a footbridge comparable, for example, with the Swanscombe footbridge over the A2 motorway.

Nevertheless in recent years some pleasing and technically interesting designs have appeared on British Railways, mostly in prestressed concrete. There have been also some attempts in the UK to achieve a measure of standardisation to enable footbridges to be assembled from standardised precast concrete units, so far, without success at the time of writing.

There is not much to be recorded about steel footbridges, and an illustration is included of the somewhat unusual Adamsdown bridge

Plate 5.1 *Adamstown Footbridge, Cardiff, B.R. Western Region*

Plate 5.2 *Congleton Footbridge, B.R. London Midland Region*

at Cardiff where steel tied arches support welded steel plate girders. Another example of a more normal type at Congleton on the electrified lines of the London Midland Region is also given.

The most interesting development in steel are those designed by Tubewright's Ltd, a subsidiary of Stewarts & Lloyds, using welded tubular construction. Two examples of these are shown; the first known as the 'Usk' type from the location of the prototype is a long bridge made up of a series of 31 m and 37 m spans at Temple Mills on the Eastern Region. The second shows a 12 m span footbridge at New Pudsey on the Eastern Region which is built up

from hollow rectangular section tubes in the form of a Vierendeel truss. There are numerous examples throughout the country of bridges of this type.

In electrified areas, or where electrification is likely, it is a requirement that a substantial parapet at least 1·5 m high be provided, and this tends towards a design of heavy appearance. If, however, the parapet is designed as a structural member with the footwalk made from precast prestressed concrete planks this can be turned to advantage. The footbridge at Scunthorpe on the Eastern Region is a striking example of this, and two pictures of it are included. It consists of five spans, excluding the ramped

Plate 5.3 *Temple Mills Footbridge: tubular 'Usk' type by Tubewrights, B.R. Eastern Region*

approaches, of 20 m and 26 m and is 183 m long. The side parapets are precast prestressed beams, and rest on precast reinforced concrete trestles 5·8 m high and 9 tons in weight. It was designed by Rendel Palmer & Tritton for Mr A K Terriss, the then Chief Civil Engineer of the Eastern Region, and was built in 1957 by Dow Mac Products Ltd.

On the Southern Region a number of footbridges built from precast concrete units were constructed in the 1930s, although the first was erected in 1926 at Pinhoe. A photograph of a rather large example of this type is shown in course of erection at Exeter, and another view taken in the maker's yard of one being

temporarily erected. These early prefabricated units tended to be very heavy, and expensive to erect, and some breakages occurred.

Some years ago the London Midland Region Civil Engineer's Department was asked to prepare a standard design of footbridge, making use of prefabricated units, and several bridges were built, but even this was not considered entirely satisfactory.

On the LM Region electrification over 100 footbridges had either to be raised, reconstructed, or entirely dismantled to enable the additional electrification clearances to be provided, and many were rebuilt in new forms. Later the tendency has been towards individual

Plate 5.4 *Pudsey Footbridge: Tubewrights hollow steel section type, B.R. Eastern Region*

designs suited to the local conditions. Several elegant and economical designs have been produced, notably the bridge at Belfast serving the Musgrave Hospital and one on the Eastern Region at York, both of which are illustrated. Due to current MOT requirements on overhead electrified lines, some heavy designs were produced in the early stages, notably No 89 on the Crewe–Stockport section of the LMR. It should be pointed out here that whereas in the case of wider road overbridges the overhead-line equipment is arranged so that the supporting structures are at equal distances on either side of the bridge, so that the catenary is at its lowest point, this is not always done in the case of footbridges. Consequently footbridges may be located at a high point on the catenary contour, and thus require a good deal of lift. This is rather exemplified by the shape of No 89 as illustrated.

Bridges Nos 23 at Runcorn, 29A at Prestbury, and 111 at Tring, all on the LMR electrified lines, are examples of the proposed new standards already referred to.

The illustration of the new footbridge at Ellesmere Port shows a design which was once considered as the prototype of a new standard design for British Railways.

The elegant design on the Eastern Region at York already mentioned was the work of the

Plate 5.5   *Scunthorpe Footbridge : prefabricated units, B.R. Eastern Region*

Plate 5.6   *Scunthorpe Footbridge : view of deck, B.R. Eastern Region*

Chief Civil Engineer's Department. The main beams and stairways are precast concrete, and the piers were constructed *in situ*. The balustrading is galvanised steel with an aluminium handrail. Conspan Ltd were the contractors and Concrete Services supplied the concrete units.

Photographs are also reproduced of Eastern Region footbridges at Bishop's Stortford and Westcliff.

The pleasing design at Belfast serves the Musgrave Park Hospital. The piers steps and haunches are cast *in situ* concrete, and the central span is built up of five separate precast reinforced concrete beams. This bridge was built by D R Martin & Sons to the design of Messrs Kirk McClure & Morton.

An interesting design at Abbey Wood on the Southern Region is illustrated. It was built for the then LCC and consists of freely supported hollow prefabricated concrete beams supported on *in situ* cast reinforced concrete piers which also carry the stairways. The piers are 7·6 m high, and the span is 21·4 m. The camber of the beams is sufficient to permit of the prestressing cables being horizontal, and these were grouted into painted metal sheaths. Mr Hubert Bennett was the architect and Howard Humphries & Sons were the consulting engineers. Holland Hannon & Cubitts were the contractors.

Two photographs of Western Region footbridges are shown. The least attractive design is that at Marazion, while the one at Tenby has a much more pleasing and contemporary appearance. It is constructed of prefabricated prestressed concrete units resting on cast *in situ* concrete trestles.

Plate 5.7   *Exeter Footbridge : concrete unit construction*

Plate 5.8 *Precast units in maker's yard*

Plate 5.9   *Musgrave Hospital Footbridge, Belfast*

Plate 5.10   *Footbridge at York, B.R. Eastern Region*

Plate 5.11 *Footbridge No 89 : raised for electrification*

Plate 5.12   *Footbridge at Prestbury: reconstructed for electrification, B.R. London Midland Region*

Plate 5.13   *Footbridge at Tring : unit construction, B.R. London Midland Region*

Plate 5.14   *Footbridge at Ellesmere Port, B.R. London Midland Region*

Plate 5.15   *Footbridge at Abbey Wood, B.R. Southern Region*

Plate 5.16 *Footbridge at Marazion, B.R. Western Region*

Plate 5.17   *Footbridge at Tenby, B.R. Western Region*

Plate 5.18 *Footbridge at Bishop's Stortford, B.R. Eastern Region*

Plate 5.19   *Footbridge at Westcliff, B.R. Southern Region*

6 Subways on British Rail

The construction of subways under railways has been revolutionised by the modern technique of 'pipe-jacking'. Previous to this they were constructed by cut and cover or by timbered excavation methods. The fact that the 'pipe-jacking' system can operate without any interruption to traffic, without disturbing the railway formation, since the void created by the tunnel is completely filled by the tunnel lining itself, and at a not unreasonable cost, is a great advantage over the old methods, especially as quite large diameter concrete pipes can now be jacked through without difficulty. This is the usual condition encountered when building a pedestrian subway alongside an existing underbridge with the tracks on an embankment, although the method is perfectly applicable to railway tracks on the level by sinking driving pits on either side of the railway to house the driving equipment. Furthermore, the rate of construction is greater.

The 'pipe-jacking' system is now well established and has already been described in Chapter 2, and for pedestrian subways concrete hollow circular sections of from 3 to 3·8 m dia are commonly used. A large number of smaller pipes for gas, water and other services with diameters from 1·5 m to 1·8 m have been so constructed by the 'Seerthrust' system operated by Wm J Rees of Woking, and fig 45 shows a diagram taken from an article by Mr Rees himself in the *Surveyor* of June 1964. Among works such as these under railways may be mentioned a 1 m dia gas main at Chislehurst, twin 1·85 m pipes for the Exeter Storm Water Drainage Scheme under the Western Region main line, a 1·8 m pipe 80 m long for Dorman Long Ltd under the Eastern Region tracks at Middlesbrough, and twin 1·5 m pipes for the Croydon Borough Council under the Southern Region tracks.

The Exeter scheme involved pipes 37 m long and 1·5 m apart. Both pipes were driven simultaneously in thirteen days' continuous driving using three 150-ton jacks.

At Middlesbrough the ground consisted of soft clay which made it necessary to drive from both sides of the railway. Trouble from water was also successfully overcome.

At Croydon the driving of the twin 1·5 m dia pipes under the tracks at Wandle Park presented greater difficulties. Prior consolidation of the ground, which was of poor quality, required the use of bentonite cement grout.

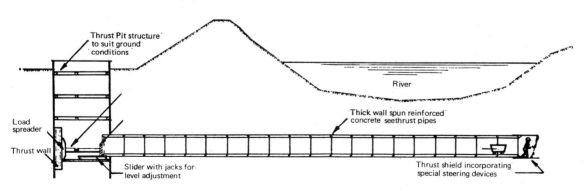

Fig 45 *'Seerthrust' pipe-jacking system*

Plate 6.1   *'Pipe-jacking': looking down into a drive pit*

1·5 m. It was considered desirable to stabilise the ground by chemical consolidation, and to reinforce the tracks by placing way beams.

Referring to fig 46, parallel return lines of Larssen No 2U sheet piles were driven to form the subway entrances at each end, and sheet pile head walls were built, item 9 at the reaction face, and item 10 at the drive face. Steel anchorages item 7 were provided at the reaction face, and the two anchorages were tied together by Lee McCall HT bars passed through 405 mm dia auger bores item 6 driven from end to end.

Chemical consolidation was then carried out in two phases, first clay/cement grout, and then silicate/bicarbonite. After this the four Lee McCall bars, two at the top, and two at the bottom, were stressed between the two anchorages. While these operations were under way the fabricated steel thrust frame item 12 at the rear of the jacking rig item 1 was constructed together with the anchor blocks 3. The HT ties were extended to the thrust wall, so that the reaction from the jacks passing through the jacking frame was carried by the HT bars through to the far face of the sheet piling at the reception end.

Driving then commenced on a 24-hour schedule until the drive was completed. The reinforced rings were 3·1 m dia, 229 mm thick, reinforced by 23 mm dia bars at 150 mm crs on the outer face, and 100 mm crs on the inner face. The recorded jacking force was 300 tons. Tube Headings Ltd were the contractors. The total length of the drive was 17 m.

Another similar subway but constructed under easier conditions was built by the same firm at Huntonbridge near Watford under the LM Region main line. Photographs of these and other pedestrian subways are included. Reference may be made to a paper on this subject read to the ICE by Mr J C Thomson in April 1967.

The Watford subway

This is a subway for vehicles, and is remarkable for two reasons; first because it is the largest hollow concrete box section hitherto driven by the 'pipe-jacking' method, and secondly instead of the tunnel being made up of a number of

Commencing at the tunnel face, an injection was made every 1·8 to 2·1 m, and 12 hours was allowed to elapse before continuing the drive, and after completion the surrounding ground was grouted as a precaution against any settlement of the permanent way. Several photographs of W J Rees' work are included.

Most of the larger sizes of subway required for pedestrians have been carried out by Tube Headings Ltd, who, jointly with British Railways, have pioneered the use of large hollow rectangular concrete sections for bridge foundations.

Fig 46 shows how a circular pedestrian subway 3·1 m dia was driven at Richmond under the tracks of the Southern Region alongside the South Circular Road. This shows how the method can be used at an embankment without the construction of driving pits. The subway is some 2·8 m away from the existing bridge abutment, and the minimum cover was only

Plate 6.2 *'Pipe-jacking': twin pipe shields, Exeter*

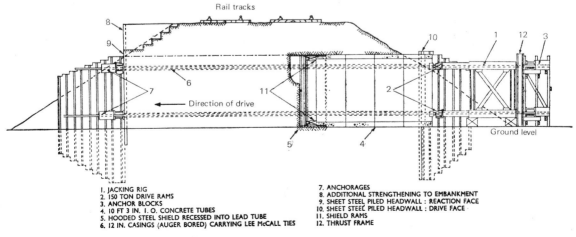

1. JACKING RIG
2. 150 TON DRIVE RAMS
3. ANCHOR BLOCKS
4. 10 FT 3 IN. I. O. CONCRETE TUBES
5. HOODED STEEL SHIELD RECESSED INTO LEAD TUBE
6. 12 IN. CASINGS (AUGER BORED) CARRYING LEE McCALL TIES

7. ANCHORAGES
8. ADDITIONAL STRENGTHENING TO EMBANKMENT
9. SHEET STEEL PILED HEADWALL : REACTION FACE
10. SHEET STEEL PILED HEADWALL : DRIVE FACE
11. SHIELD RAMS
12. THRUST FRAME

Fig. 46 *Richmond pedestrian subway*

Plate 6.3 *Pedestrian subway at Ash, B.R. Southern Region*

relatively short lengths, it was driven in two half-lengths of 18·7 m and 21·8 m, the longer length being over 500 tons in weight. The external dimensions of the tunnel were 8·3 m by 3·7 m.

The tunnel was required to provide access for vehicles from Watford Station forecourt to a car-parking site which had been created on the opposite side of the Euston to Scotland main line.

The reason for constructing this tunnel in two half-lengths was the limited space available in the station forecourt, which would have involved difficulties with crane working, transport and storage of shorter individual

units, apart from their great weight and cross-sectional size. The building of the tunnel half-unit could be confined to a corner of the forecourt without causing serious obstruction to the existing road and pedestrian facilities to the station itself. It had been estimated that the driving of such a large concrete unit would require a line possession of about seventy hours or one week-end.

The two half-units were built on either side of the route, and were driven by separate jacking systems to meet in the centre of the railway.

Fig 47 shows a longitudinal section through the subway. At the north, or car-park, side

Plate 6.4 *Watford vehicle subway: jack assembly, B.R. London Midland Region*

beams to give jacking resistance as the units moved forward.

Each unit was provided with a massive driving shield with steel cutting edges, and during the drive bentonite slurry was injected along the sides and roof to reduce friction.

The two driving shields were not recovered but were left *in situ* and were concreted in and a ventilating shaft constructed above them. The external dimensions of the tunnel shown on fig 48 are 8·3 m by 3·7 m, and internally 7·4 m, by 2·8 m. The walls, floor and roof are 45·7 cm thick except the south side wall which is 53·7 cm thick and incorporates a drain. The tunnel contains a two-lane roadway 5·6 m wide, and a footpath 1·24 m wide. The headroom above the road surface is 2·5 m. As the headroom below rail level was only 90 cm, it was considered prudent to close the two tracks involved by the driving of each unit and permitting traffic

there were no unusual problems apart from the size and length of the unit, because being on high ground the usual sheet pile walled driving pit could be built, and the ground provided a sufficient abutment to take the thrust of the jacks. On the south side the bottom of the tunnel was practically at ground level, and it was necessary to devise other methods of providing jacking resistance.

Both half-units were built on a massive reinforced concrete base with a relatively low thrust wall at the driving end, since it was considered that with such heavy units it would only be necessary to apply the thrust at the bottom. To resist the thrust of the jacks the base at the south side was provided with twenty inclined ground anchors which penetrated 12 m into the ground at an angle of 45°.

Each base was also provided with rows of sockets 1·5 m apart into which heavy jacking pins were inserted to give anchorage for strong

Plate 6.5 *Watford vehicle subway: driving operations, B.R. London Midland Region*

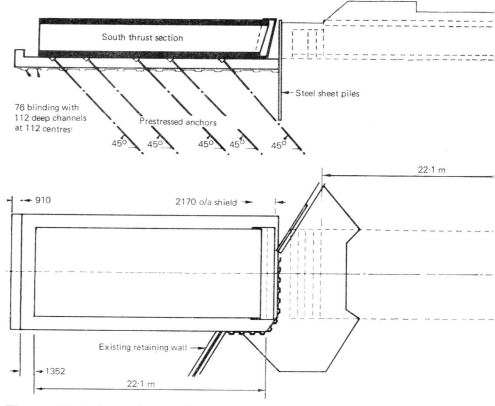

Fig 47　*Watford car subway: plan and longitudinal section, B.R. London Midland Region*

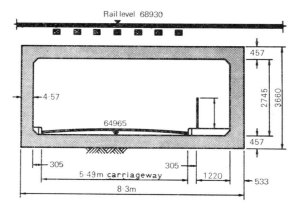

Fig 48　*Watford car subway: cross-section, B.R. London Midland Region*

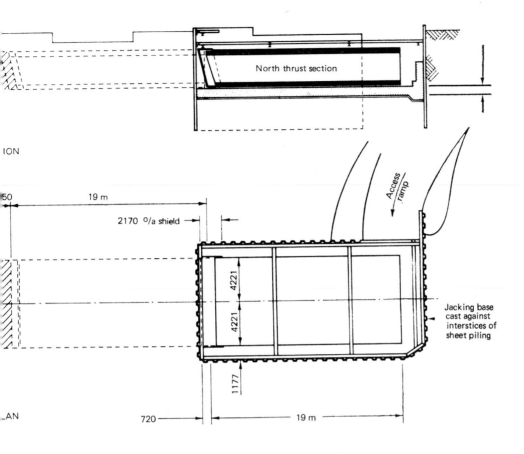

North thrust section

ION

50   19 m

2170 o/a shield

Access ramp

4221

4221

1177

Jacking base
cast against
interstices of
sheet piling

AN

720   19 m

to use the other two tracks of the four track route. One unit was therefore driven during the week-end April 25/27, 1970, and the second on 23/25 May. Actually some disturbance of the shallow ground above the tunnel took place due to low cohesion, but this was easily rectified. The approaches to the tunnel were constructed *in situ* at each end. Ten 150-ton jacks were used.

Tube Headings Ltd carried out the driving and mining work. All other civil engineering work was done by the London Midland Region Direct Labour Organisation, including the casting of the tunnel units on their bases. The whole scheme was under the overall supervision of Mr W F Beatty, Chief Civil Engineer, LM Region, British Railways.

7 *Some notable bridges overseas*

CENTRAL AFRICA

River Nile bridge at Pakwash

This bridge over the Albert Nile completed in mid 1969 makes the first rail link between the Congo and Uganda, and is located some 24 km north of Lake Albert. Hitherto the nearest bridge over the Nile north of Lake Albert had been the Sudan Railway's bridge at Kosti, 1,280 km away. It has enabled the East African Railway's extension northwards from Soroti to be continued further. It is therefore of some importance, although not a large bridge as compared with many others in Africa.

It consists of three simply supported Warren truss spans of 81 m, and carries a 7 m wide roadway with a single-track metre-gauge railway in the centre. It is a typical design to be met with all over the African Continent. The design loading for the road was BS 153 and for the railway fifteen units RA2. The trusses are of normal design, principally in high tensile steel with a reinforced concrete deck acting compositely. The steelwork was fabricated in Britain, and the site joints were made with Torshear high strength friction grip bolts. This enabled speedy erection by local labour to be achieved under supervision by the steel-work contractor.

The two piers and abutments are founded on 138 cm dia bored reinforced concrete piles sunk to a minimum depth below ground level of 15 m and up to 30 m at the piers.

A somewhat unusual problem during foundation construction arose from large floating masses of papyrus grass varying in area from 8,000 to 12,000 m² which were large enough to create what was in effect a dam when encountering an obstruction such as a bridge pier, and were thus a source of danger. To counter this, large reinforced concrete fender blocks founded on three bored piles were provided on the upstream side of the piers. These can be seen on the photograph of the bridge reproduced.

For the construction of the bridge use was made of a disused floating dock from Lake Albert not only to sink the piles but to erect the trusses which were floated into position. This operation occupied from three to five hours.

The contractors were Mowlem Construction Ltd of Nairobi, and the consulting engineers for the design of the substructures were Messrs Goode & Partners of London.

The Crown Agents for Overseas Governments and Administrations were responsible for the superstructure design.

Dorman Long (Bridges & Engineering) Ltd were the subcontractors for the steelwork, which was fabricated by the Cleveland Bridge & Engineering Co Ltd.

CHINA

River Yangtze Bridge at Nanking

A fine example of railway-bridge engineering in China is the ten-span bridge over the Yangtze River at Nanking, opened in 1968, twelve years after the completion of a similar bridge at Wuhan over the same river but higher up. It provides a direct rail route between Shanghai and Peking. Site plan fig 49.

Plate 7.1 *River Nile Bridge at Pakwash, Uganda*

Fig. 49   *Bridge over the Yangtze River : site plan*

The bridge is 1600 m long, and is a double-deck cantilever steel structure to carry both rail and road traffic. The dual-carriageway road is on the upper deck and the double-track railway on the lower deck. The distance between the two decks is 20 m.

The river at the point of crossing is 1,200 m wide, and the bed consists of a thick layer of alluvium on broken and flawed rock. Solid rock lies at a considerable depth. The nine masonry piers are 130 m apart and were constructed in coffer dams. While the bridge itself is 1,300 m long, the overall length, including approaches, is 6,700 m.

Chemical coatings were used for the protection of the steelwork.

While the first bridge built in 1956 at Wuhan was done with Russian assistance, this one was designed and built entirely by the Chinese. The steel was made and rolled at the Anshan Iron and Steel Works in Manchuria, and fabricated and assembled at the Shanhaikwan Bridge Works of the former Peking–Mukden Railway. The photograph reproduced of the inaugural train indicates the pride of the Chinese People's Republic in this fine work.

INDIA

Reconstructed bridge over the River Juggal

Although this is a relatively unimportant bridge on the 170 km 77 cm gauge Kangra Valley Railway in the Himalayan foothills, its repair after a washaway on the Juggal River is of interest. When it was built in 1928 its construction presented several engineering problems to the engineer, Colonel Everall. It was built to serve a hydro-electric project and is mostly laid on a shelf cut out of the gorge of the River Banganga.

As can be seen on the site plan, fig 50, the bridge crosses over the River Juggal at its junction with the Banganga, and it is the sweep of the Juggal round the Pathancot bank which has caused the erosion of the base of the Juggal gorge and the undermining of the bridge support after the lapse of thirty years. The cliff here is 77 m high.

The original bridge consisted of three 12·4 m deck type steel girder spans resting on masonry piers, and a final 31 m Warren triplicate underslung truss span resting on an abutment cut out of the rock of the cliff on which was constructed a concrete raft surmounted by a massive bedstone seating. When the erosion of the cliff amounted to some 2·4 m the position became serious, and, in fact the cliff collapsed.

The condition is shown at A on the diagram, fig 51, and the succeeding diagrams show the steps which were taken to restore the bridge by inserting a new 13·6 m plate girder span resting on the last pier and a new pier, and moving the Warren truss span 14 m towards Pathancot on to a new seating cut out of the cliff. First of all, temporary trestling was erected close to the cliff face as shown at B, and while the new pier was being built traffic at reduced speed was maintained. The new plate girder span was then brought along the existing track on trolleys until it rested as shown at B. Work on excavating for the new seating was also carried on, making use of shallow-service girder between the end of the truss and the new seating, to maintain traffic. The bridge was then closed for the final operations.

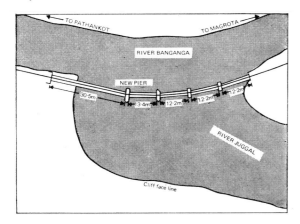

Fig 50   *River Juggal Bridge : site plan*

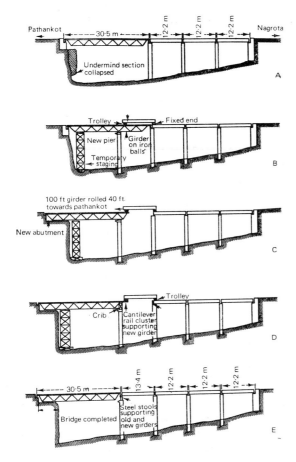

Fig 51   *River Juggal Bridge : method of reconstruction*

The old span was then moved on rollers from underneath the new plate girder until it reached the position shown at C and D on fig 51. The new plate girder was then lowered on to its seatings as shown at E.

The operation was made more difficult by reason of the curvature of the alignment which required the new span to be mitred to fit into its position. Also the monsoon rains were approaching, and in fact rain and floods struck the bridge within hours of the reopening to traffic and the temporary staging was washed away completely. Photographs reproduced show the original collapse of the cliff, and the final operation of moving the old span and the insertion of the new one.

PAKISTAN

The Attock Bridge over the River Indus

The original bridge, located about 3 km below the junction of the Indus and Kabul rivers, was built in 1883. In 1921 not only was it desired to strengthen it to take heavier traffic, but site investigation revealed considerable deterioration of the steelwork, and the central pier in midstream, and it was decided to rebuild.

Although it is now forty years since the rebuilding took place it is still a remarkable structure, and the method of rebuilding was noteworthy. Messrs Rendel Palmer & Tritton were the consulting engineers, and by their kindness I am able to show two photographs. The reconstruction was very fully described in a paper read to the ICE by Mr W T Everall in April 1930 from which the brief notes which follow are taken. The drawing, fig 53, shows the original bridge, the reconstruction stage, and the final form. The River Indus is subject to considerable and severe flooding during the monsoon months, and it is so happened that in 1926, when reconstruction commenced, the second worst recorded floods occurred and the river level rose by 17 m. A highest recorded level was in 1929 when the waters rose 20·5 m. The work was therefore carried out under considerable difficulties.

Plate 7.2 *River Yangtze Bridge, China:*
general view

Plate 7.3 *River Yangtze Bridge, China:*
opening ceremony

Plate 7.4   *River Juggal Bridge, India : bank collapse*

Plate 7.5   *River Juggal Bridge, India : placing the new girder*

The original bridge, 131 m long, consisted of five spans: two central spans over the river of 94 m each; two approach spans of 78 m at the Rawalpindi end; and one also of 78 m at the Peshawar end. The bridge carried a roadway and a single-track railway, the road being at the bottom of the steel girders and the railway on the top. The girders were double-intersection lattice type simply supported on wrought-iron trestle piers and masonry abutments. The central pier was founded on a rock island in midstream. As it was required to make the bridge double track, the new girders had in any case to be larger, so it was decided to build the new girders round the old ones and use the old girders to carry 25-ton travelling cranes which performed the erection. This is clearly shown on fig 53, and in the photograph. Erection was by cantilevering with wedge gear for adjustment at temporary supports. The approach spans were strengthened by converting them to continuous girders by the

Plate 7.6 *Attock Bridge, River Indus, Pakistan*

addition of an extra pier as well as by strengthening the steelwork itself to carry the heavier loading. The girders in the approach spans are similar to the original ones, but the two central girders are curved lower chord type which gives the bridge its rather striking appearance as the illustration shows.

BURMA

The Sittang River bridge

The original bridge over the River Sittang was built in 1908, and was destroyed during the second war. It consisted of eleven spans of 46·5 m, and provided the only link between Rangoon, Pegu, and Martaban. It carried a single metre gauge railway and a footwalk.

The turbulence of the River Sittang and its tidal bore, which caused a rise in water level of 2·8 m in fifteen minutes, had made necessary a lot of protective work over the years, so that the idea of building a new bridge had been

envisaged before 1939. Fig 52 shows a site plan of the area and the location of the new bridge some 7 km upstream.

Due to the disturbed conditions in Burma

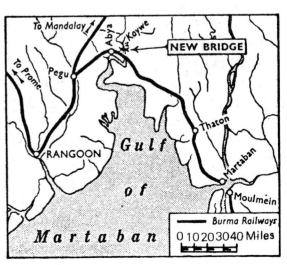

Fig 52 *Sittang Bridge : site plan*

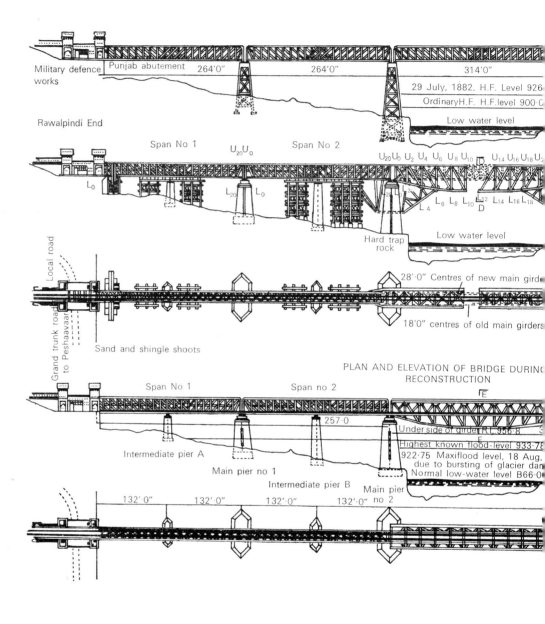

Military defence works

Punjab abutement   264'0"   264'0"   314'0"

Rawalpindi End

29 July, 1882, H.F. Level 926
Ordinary H.F. H.F. level 900·0
Low water level

Span No 1 $U_{20}U_0$ Span No 2 $U_{20}U_0$ U_2 U_4 U_6 U_8 U_{10} D U_{14} U_{16} U_{18} U_2

L_0 L_{20} L_0 L_2 L_4 L_6 L_8 L_{10} L_{12} L_{14} L_{16} L_{18} D

Hard trap rock

Low water level

Local road

Grand trunk road to Peshaavaar

Sand and shingle shoots

28'0" Centres of new main girder
18'0" centres of old main girders

PLAN AND ELEVATION OF BRIDGE DURING RECONSTRUCTION

Span No 1   Span no 2   257·0

E
Under side of girder R.L. 956·8
Highest known flood-level 933·78
922·75 Maxiflood level, 18 Aug, due to bursting of glacier dam
Normal low-water level B66·0

Intermediate pier A

Main pier no 1

Intermediate pier B   Main pier no 2

132'0"   132'0"   132'0"   132'0"

PLAN AND ELEVATION O

Fig 53 *Attock Bridge : elevation and plan*

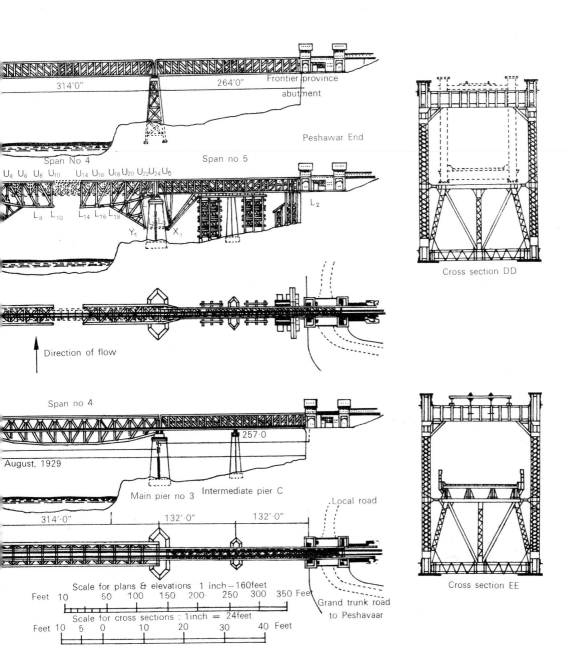

314'0" 264'0"

Frontier province
abutment

Peshawar End

Span No 4 Span no 5

U_4 U_6 U_8 U_{10} U_{14} U_{16} U_{18} U_{20} U_{22} U_{24} U_0

L_8 L_{10} L_{14} L_{16} L_{18} L_2

Y_1 X_1

Cross section DD

Direction of flow

Span no 4

257·0

August, 1929

Main pier no 3 Intermediate pier C

Local road

314'·0" 132'·0" 132'·0"

Cross section EE

Scale for plans & elevations 1 inch—160feet

Feet 10 50 100 150 200 250 300 350 Feet

Scale for cross sections : 1inch = 24feet

Grand trunk road
to Peshavaar

Feet 10 5 0 10 20 30 40 Feet

IDGE AS RECONSTRUCTED

Plate 7.7 *Sittang River Bridge, Burma*

after the war it was not until 1957 that work on the new bridge could commence. It is 716 m long and consists of six steel truss spans which carry two roadways with a single-track metre gauge railway in between, and two footwalks outside the main girders. At the west end there are two 18·6 m plate girder approach spans, of the half-through type. The two roadways are 3·7 m wide, and the footwalks are 0·93 m wide carried on cantilever extensions to the bridge cross-girders.

The construction of the seven piers was accomplished by sinking seven large steel caissons into the river bed, the deepest being 35 m below bed level. The sinking operations were conducted from fixed platforms 31 m × 37 m resting on 148 steel pipe piles driven 28 m into the river bed.

The caissons were of double-Dee construction, 15 m long by 9 m wide. The wells were sunk by open dredging in two 4·6 m dia dredging holes. The steining of the wells and piers was in reinforced concrete with precast concrete blocks for shuttering.

The truss spans are of the curved upper chord type with vertical K bracing. They were designed by Rendel Palmer & Tritton, and fabricated in Yugoslavia under their supervision and shipped to Burma. They are 111·8 m long and 16 m deep at the centre, and weigh 793 tons. They were erected near the site on special staging and floated on the rising tide upstream to the bridge site on pontoons 33·6 m by 22 m by 1·7 m where they were placed in position on the pier seatings. Preliminary construction work commenced in 1957 and was completed in 1959 and considerable difficulties were encountered not only from the river itself, but from sporadic fighting by rebel groups against the Government. No work was possible during the monsoon. Construction work proper began in 1960, and continued until 1961. It was formally opened in March of that year. The staff were entirely Burmese with a British engineer adviser.

JAPAN

The River Arakawa bridge

The prestressed concrete portions of this composite steel and concrete bridge present some features of interest.

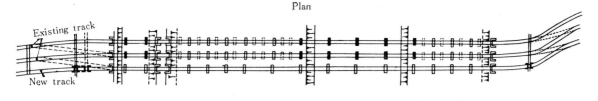

Plan

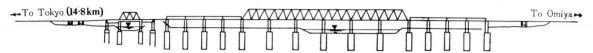

Elevation

Fig. 54 *Arakawa Bridge : elevation and plan*

Plate 7.8 *Arakawa River Bridge, Japan*

It is located about 11 km from Tokyo on the northern main line of the Japanese National Railways and comprises six tracks. These cross the River Arakawa in pairs on three parallel identical bridges all resting on common piers and abutments.

Each bridge is made up of four approach prestressed concrete spans 39 m long, a central steel truss bridge 372 m long, and four more 39 m long concrete spans identical with the others, making a total length of 685 m.

The central steel section is a continuous Warren truss of normal construction in six spans. It is the concrete portions which are of particular interest. Fig 54 is a diagram of the plan and elevation of the bridge.

Fig 55 shows a cross-section of the prestressed concrete approach spans. They are through slab spans with two side girders splayed outwards. This is to give adequate clearances to the trains. Normal JNR practice with bridges of this type, whether in steel or concrete, is to use three girders. It was, however, necessary to maintain the normal track centres of 4·18 m standard for the 3 ft 6 in (1·1 m) gauge, and when the design was worked out it was found

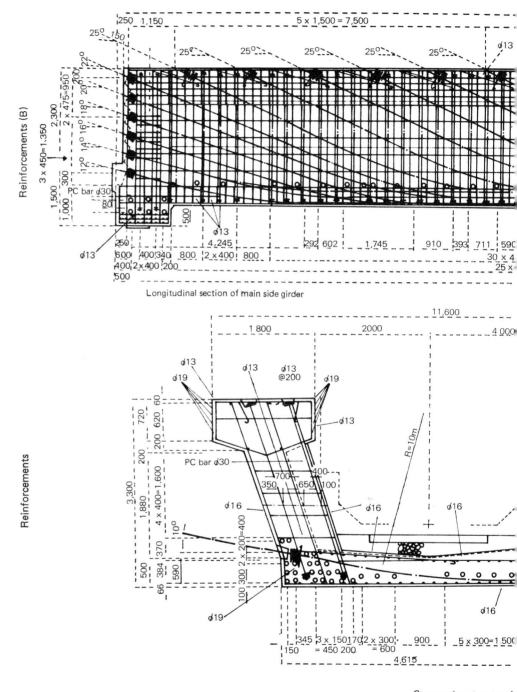

Reinforcements (B)

Longitudinal section of main side girder

Reinforcements

Cross section at centre of

Fig 55 *Arakawa Bridge : section*

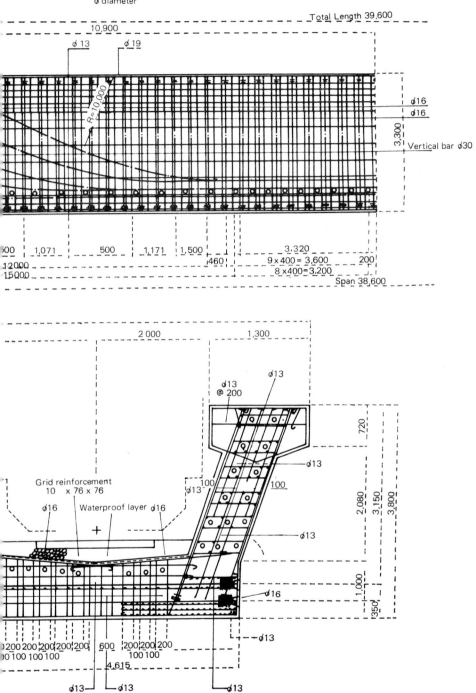

ϕ diameter

Total Length 39,600

10,900

ϕ 13 ϕ 19

R=10,000

ϕ16
ϕ16
3,300
Vertical bar ϕ30

500   1,071   500   1,171   1,500   3,320
                          460   9 x 400 = 3,600   200
12000   8 x400=3,200
15000   Span 38,600

2 000   1,300

ϕ13
ϕ13
@ 200 ϕ13

720

ϕ13
100   ϕ13   100
2,080   3,150   3,800

Grid reinforcement
10 x 76 x 76
ϕ16   Waterproof layer ϕ16 ϕ13

ϕ13

1,000

ϕ16

350

ϕ13

200 200 200 200 200   600   200 200 200
100 100 100 100   100 100
4,615

ϕ13 ϕ13 ϕ13

Cross section at end support

REINFORCEMENT DETAILS

Plate 7.9   *Arakawa River Bridge, Japan : end view*

that there was insufficient space for the top flange of the centre girder. Further, the approach gradient of 1%, and the need to allow maximum possible clearance for floods, limited the depth from rail level to the underside of the floor to 1·1 m and made this design the most suitable. Actually designs in steel plate girders, and a through steel truss were investigated before the prestressed concrete design was finally adopted. Such a design presented many problems particularly torsional ones, and tests of an exhaustive character were made on a model one quarter full size made with a special light aggregate concrete.

In the actual bridge there are fourteen main prestressing cables in each of the main girders, and thirty-two longitudinal prestressing cables in the slab, together with a large number of transverse prestressing cables. The tests on the scale model showed that ordinary beam theory could be applied on the assumption that the deck slab formed the tension flange of the two side girders, and that Jacobsen's methods could be adapted and modified to give practical and realistic values for torsional stresses.

A full account of the work on these bridges is to be found in Report No 30, Vol 9, No 1, of 1966 of the Permanent Way Society of Japan.

Rubber bearings form the seatings for the bridges, on which the maximum bearing stress is 60 kg/cm².

AUSTRIA

Two bridge renewals on the Arlberg line

In November 1968 two important bridges were installed on the Innsbruck–Bludenz section of the Arlberg main line of the Austrian Federal Railways, one rebuilt as a result of an avalanche, and the other due to renewal after war damage.

The Schanatobel Bridge, 42 m long, was lifted off its bearings and destroyed in January 1968 after two simultaneous avalanches. A temporary bridge was installed which enabled traffic at reduced speed and weight restriction to be maintained.

A new steel tied arch bridge was selected as a replacement and this was erected at the side of the temporary bridge in such a manner that the temporary bridge could be moved out on temporary steel supports and the new one moved into position in one operation. New heavy anchorages were installed to prevent a recurrence of the damage by the avalanche, and the gully above the bridge was widened by the removal of 5,000 m³ of rock as a further

Plate 7.10   *Schnatobel Bridge renewal, Austria*

Plate 7.11   *Otztaler Bridge renewal, Austria*

precaution. The new bridge weighs 155 tons and the parapets are low to facilitate snow clearance. The line is single track.

Photographs show the temporary bridge and the new bridge being erected side by side, with the new bridge in position and the old ready to be removed.

The other bridge, some 80 km to the east on the same line, carried the line over Otztaler Ache, and had a main span of 60 m, with two

Plate 7.12   *River Tagus Bridge, Portugal*

20 m spans at one end and one 20 m span at the other. It had been heavily damaged during the war, and one 20 m span had been renewed, and the rest of the bridge patched up by welding.

The two patched spans have now been replaced by a continuous girder with one 60 m and one 40 m span. It is a steel box girder 3 m deep and 2 m wide, and replaces the original lattice steel main span. The greater depth of the new bridge necessitated alterations to the tops of the piers and abutments, and advantage was taken to make them suitable for an additional track at a future date.

The new bridge was erected on the ground at

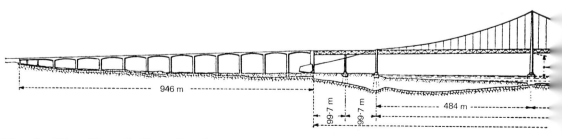

Fig 56   *River Tagus Bridge : elevation*

one end of the gorge, and then moved alongside the old one on to falsework, so that both could be moved out and in on rollers in one operation similar to the way adopted for the Schnatobel Bridge. In fact, both bridges were moved on the same date in November 1968.

In the case of the Otztaler Bridge the old truss span was pushed bodily over into the gorge below.

Both bridges are examples of modern design and construction. Photograph shows the Otztaler Bridge after rolling in, with the old truss span lying below.

PORTUGAL

The River Tagus suspension bridge at Lisbon

Large suspension bridges are always interesting and striking and comparatively few are involved with railways. The suspension bridge over the River Tagus at Lisbon is remarkable not only for the above reasons but also because at the time it was built it had the longest main span in the world—1,030 m and the deepest pier foundation—80 m. The continuous truss forming the deck was also at that time the longest—2,325 m. An outline of the bridge is shown in fig 56.

The bridge is designed for a four-lane road carried on the top deck, and a double-track railway on the lower deck.

It is of the stiffened suspension type, and on each side of the main 1,030 m span there is an approach span of 492 m, and the overall distance between the anchorages is 2,316 m. The clear headway over the water for shipping is 71 m.

On the south side the bridge starts at high ground, so that the cable foundation is formed in solid rock. On the north or Lisbon side the cable foundation is at ground level some distance below bridge level, and forms part of the column supporting the 202 m extension of the suspended girder beyond the north pier. From the end of the suspension bridge proper extends a long viaduct for 961 m in prestressed concrete construction over the city with eleven high reinforced concrete columns spaced from 75 m to 91 m apart. This in itself is a major engineering work.

The deepest foundation is at the south pier. Both pier foundations were constructed by the open caisson method. The caissons were rectangular in plan, the lower part being in steel with the usual cutting edge, and the steining in reinforced concrete. The caisson for the south pier is 31 m × 24 m wide and contains twenty-eight dredging wells 4·8 m dia. The whole foundation is 80 m deep.

The north foundation which is shallower, is 31 m × 18 m wide, and has twenty-one wells. They were launched from a construction site some 5 km away, and floated away to the bridge site where they were sunk by pouring concrete round the wells and caisson walls.

The two towers carrying the suspension cables are of cellular steel construction 194 m high with legs splayed sufficiently to allow the continuous deck to pass between them.

The girders of the deck are of the Warren-truss type, fabricated in special high strength steel of 733 to 924 kg/cm² tensile strength. Shop connections in the towers were riveted,

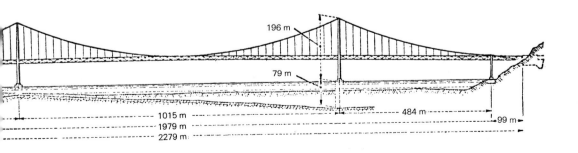

but those in the decks were welded, and all main field connections were made with high tensile bolts.

There are two suspension cables spaced 24 m apart, and each consists of thirty-seven strands each made up of 304 galvanised high strength steel wires 4·8 mm dia, giving an overall dia of 525 mm.

The anchorages are capable of withstanding a load of 25,000 tons. On the south side the approaches required a cutting through the rock involving the excavation of 2,523,200 m³ of earth and rock. The bridge was started in 1963 and completed in 1966.

The main contractor was the United States Steel International Inc, working in collaboration with the Civil Engineering Department of the Portuguese Government.

FRANCE

Rail connections to Rungis market

When the old perishable food market at Les Halles in the centre of Paris was moved out into the suburbs at Rungis, near Orly, new rail connections were required which involved one or two modern-type bridges.

One such steel bridge, No C14, built by the SNCF, carries the shunting neck between the reception sidings group and the Rungis market proper across a main road. Also on it is located the shunting 'hump' and control cabin over which the freight vehicles are pushed to move by gravity into the market sidings.

A diagram, fig 57, shows the elevation and plan of the bridge which crosses the Route Nationale 186. It is 89 m long and 25 m wide and is composed of three spans of 21·5, 31, and 26·5 m. It is a composite steel and concrete structure, having a reinforced concrete deck resting on a welded steel plate substructure which is, in effect, a cellular steel box section mattress. It is an interesting form of construction. The concrete deck is bonded to the steel by means of steel hoops welded to the top plates of the substructure. On fig 58 are given cross-sections which show how the elevation of the hump was achieved. The welded and bolted construction of the bridge is quite a complex one.

The bridge has a fixed bearing on the northern abutment, roller bearings on the two piers, and a free bearing on the southern abutment. The five steel box girders are 5 m apart and were fabricated in units which were bolted together.

Plate 7.13　*Bridge No C14, France*

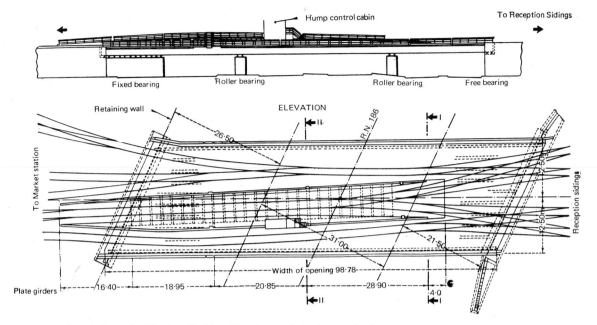

Fig 57 *Paris Rungis Market Bridge No C14 : elevation and plan*

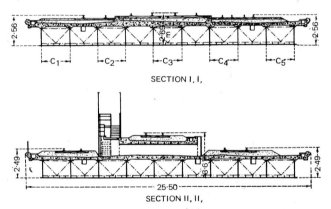

Fig 58 *Paris Rungis Market Bridge No C14 : cross-section*

The two abutments are of masonry construction resting directly on the subsoil, and in their design account was taken of the additional loadings due to the braking of vehicles passing over the hump. The two reinforced concrete piers each consist of five separate pillars resting on a common concrete foundation which was constructed within sheet pile walls. The minimum deck thickness is 2·39 m. An enlarged detail of the steel construction is shown in fig 59.

Viaduct on rail connection to Rungis Market

One of the new rail connections required was a single line from Juvisy marshalling yard to the new market, and this involved a fine reinforced concrete viaduct across the town of Orly. A photograph is reproduced. It is a three-

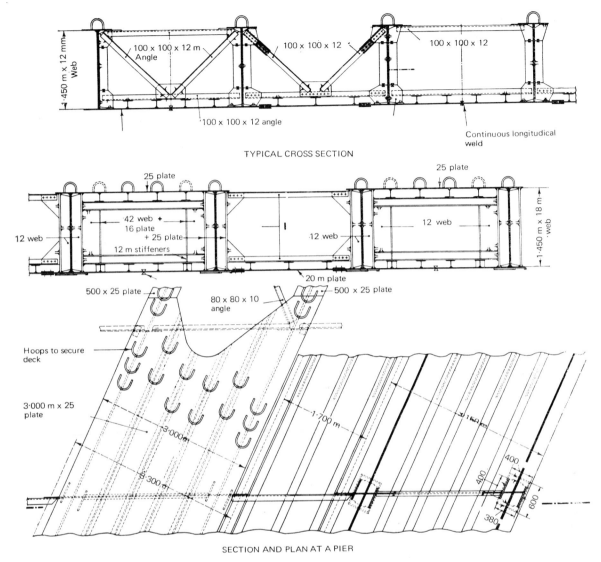

TYPICAL CROSS SECTION

100 x 100 x 12 m Angle

100 x 100 x 12

100 x 100 x 12

1·450 m x 12 mm Web

100 x 100 x 12 angle

Continuous longitudical weld

25 plate

25 plate

12 web

42 web + 16 plate + 25 plate

12 m stiffeners

12 web

12 web

1·450 m x 18 m web

20 m plate

500 x 25 plate

80 x 80 x 10 angle

500 x 25 plate

Hoops to secure deck

3·000 m x 25 plate

3·000 m

3·000 m

1·700 m

160 m

400

400

600

380

SECTION AND PLAN AT A PIER

Fig. 59   *Paris Rungis Market Bridge No C14 : details*

span concrete viaduct 91 m long on a curved alingment and is composed of a hollow rectangular section reinforced prestressed concrete beam with a reinforced concrete deck cast integrally with it. Except for the fact that the section is rectangular and not trapezoidal, the design is reminiscent of one of the Penrith bridges on the London Midland Region of British Railways.

The main beam is 3·2 m wide; 2·5 m deep, with sides 40 cm thick. The deck is 5·4 m wide.

The reinforced concrete abutments and piers rest on groups of piles, with additional groups of raking piles at the abutments. The pile groups are capped with reinforced concrete blocks. At the Juvisy end the viaduct rests on a fixed bearing, and on roller bearings at the Rungis end. The three piers are tapered in transverse elevation, and in fact are hinged props. Both abutment bearings have lateral restraint. The minimum headroom over the road is 7·4 m. It was designed and built

Plate 7.14   *Bridge No CF6, France*

Plate 7.15   *Caronte swing bridge reconstruction, France*

by the Civil Engineering Department of the SNCF.

The Caronte viaduct and swing bridge

Located at Port de Boue, on the line between Miramos and l'Estaque, the original swing bridge in the centre of the viaduct crossing the Caronte Canal was destroyed by the Germans in 1944.

It was replaced in 1948 by a temporary single-track vertical-lift bridge over the canal.

In 1954 the swing bridge was replaced as

Plate 7.16   *Caronte swing bridge completed, France*

Plate 7.17   *Kleinenbelt Bridge, Denmark*

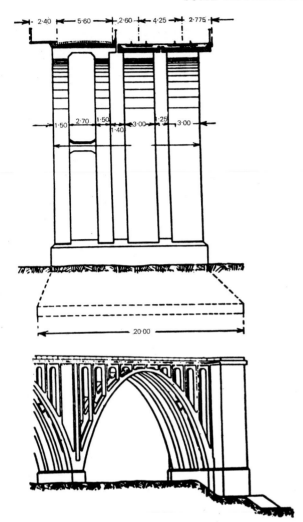

Fig 60 Kleinenbelt Bridge and Viaduct, Denmark: detail of approach span

shown in the photograph. It is interesting as being one of the longest swing bridges in the world, having a mean length of 94 m, and in the method of construction, with the temporary single-track lift bridge still in operation. This was done by building the rotating portion on top of its reinforced concrete tower parallel with the canal, and leaving sufficient space in the centre section for the passage of trains. The canal is 42 m wide, and the clear headroom above the water is 23 m.

The viaduct consists of two approach bays,

each of 51·2 m, the swinging portion with a total length of 113 m, and eight further bays of 82·5 m each.

The support for the swing bridge is a hollow reinforced concrete circular tower faced with Cassis stone, and the bridge rotates on a massive roller bearing with a ring of sixty-three steel rollers each 500 mm in diameter. It requires 50 hp to operate, which is performed by electric motors.

It is fully automatic with the most modern locking devices. It weighs some 1,200 tons.

DENMARK

The Kleinenbelt Bridge

Although built some thirty-five years ago, this 1,128 m long bridge over the Kleinenbelt is still a remarkable structure, and in 1963–4 it was strengthened by the addition of sections of broad flange beams and flat plate stiffeners.

It is shown in fig 61 and in an aerial photograph.

It is a combined rail and road bridge and connects the Danish island of Funen with Jutland, and thus provides an important link between the port of Esbjerg and Scandinavia and Hamburg and north Europe.

It consists of five steel-truss spans with a total length of 825 m with reinforced concrete arch approach viaducts at each end. There are three arches at the Funen end, totalling 138 m, and five at the west end, totalling 214 m. The arches are in four sections, two supporting the 5·6 m wide roadway, and two supporting the rail tracks. These latter are 3 m wide, and the two arches for the road are 1·5 m wide. The road and rail bridges, though adjacent, are independent.

The piers are 72 m high and are founded on well-type foundations. Details are shown on the drawing.

A GROUP OF SWISS BRIDGES

In making a selection of interesting bridges in countries like Switzerland or Austria it is slightly difficult to avoid giving undue emphasis to

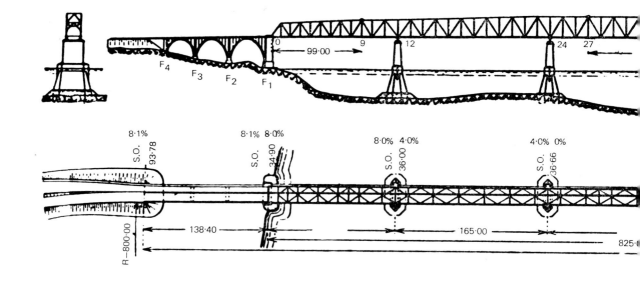

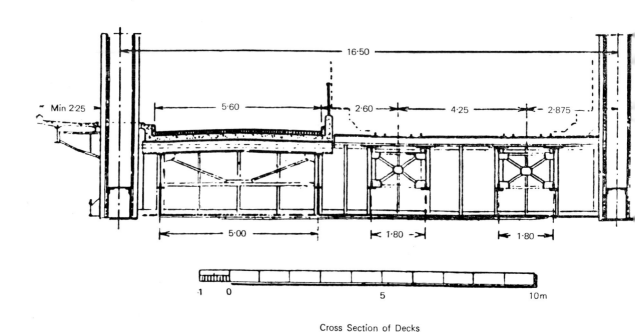

Cross Section of Decks

Fig 61   *Kleinenbelt Bridge and Viaduct, Denmark*

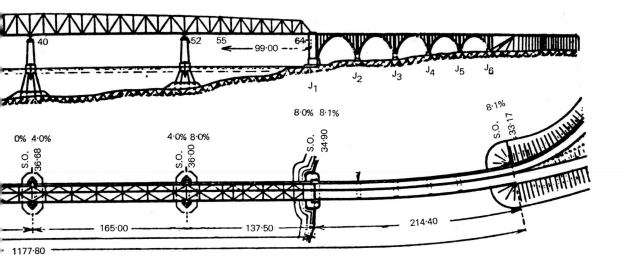

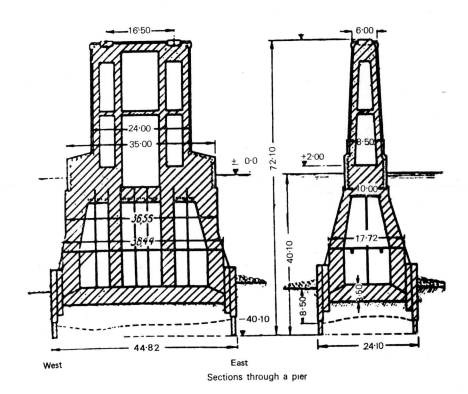

West   East

Sections through a pier

Plate 7.18 *Kander Viaduct, Switzerland*

scenic beauty at the expense of engineering interest.

By the courtesy of the Swiss Federal Railway authorities a group of seven photographs of typical Swiss bridge engineering are given.

Three viaducts, the Wander, the Luogelkin, and the Landwasser, are fine examples of masonry arch construction.

The Gstaad Viaduct is a normal three-span lattice steel girder bridge on masonry piers. The rather striking steel arch bridge at Bietschtal near Valais was designed for future double-track widening, as can be seen in the photograph.

Lastly two examples of concrete design are given. The six-span concrete arch viaduct—

Grandfey—near Friberg is supported on vertical columns resting on the bridge arches springing out of the concrete piers.

The fine modern prestressed concrete single-arch bridge—the Lorraine Viaduct—is reminiscent in appearance of the road bridges at Taf Fechan and Nant Hiz on the Heads of the Valley Road in South Wales designed by Rendel Palmer & Tritton.

As would be expected in Switzerland, the Swiss engineers are abreast of the times in all modern techniques of bridge building and erection. The mountainous character of the terrain is sometimes of assistance, in that the use of suspension cables to facilitate the erection of steelwork and placing of materials is possible.

Plate 7.19   *Luogelkin Viaduct, Switzerland*

GERMANY

The Fehmarndsund Bridge and Viaduct

This fine steel viaduct and bridge built in 1963, makes possible a direct rail link between Hamburg in Germany and Copenhagen in Denmark. It consists of eight spans with a total length of 963 m, made up of seven approach spans of 102 m each, and a steel arch navigational span of 249 m. There are five approach spans on the Danish side and two on the German side.

The sixth or navigational span is of the 'tied arch' type, from which the steel deck is suspended by steel tension bars, and is of interest, by reason of the fact that it was the prototype of the Frodsham bridges on the LM Region of British Railways described in Chapter 4. There are, however, some important differences, the principal one—apart from the much greater size 249 m as against 64 m span—being that the British one is in prestressed concrete, while the German is in welded steel.

It is a combined road and rail bridge, carrying a single track railway, and a roadway 11 m wide, with a footwalk on either side. The approach spans consist of twin hollow steel box section plate girders combined with a welded steel cellular deck with an overall width of 20·95 m. The two main longitudinal girders are 3·4 m deep, and 2·4 m wide, and one is located under the rail track, and the other under the centre of the roadway. The two are cross connected at

Plate 7.20   *Landwasser Viaduct, Switzerland*

Plate 7.21   *Gstaad Viaduct, Switzerland*

Plate 7.22   *Bieschtel Bridge, Switzerland*

intervals by bolted steel lattice work, and are 11 m apart. The construction is shown in fig 61. Each span is simply supported on cellular construction reinforced concrete piers, the largest of which are the two which support span No 6.

Fig 63 shows sectional elevations of one of the two large piers, and one of the smaller ones. They were constructed in coffer dams, and were sunk into the sea bed to a depth of 15·2 m below high water mark. The large pier has a base 37 m × 16 m × 8·7 m thick. The pier column erected on top of the base is of cellular construction, and is tapered in transverse elevation with a width at the base of 29 m and at the top 34 m. In side elevation it is 8·5 m wide. The total height from base to bearing level is 35·5 m, which gives a clear headroom for ships of some 20 m. The other piers are of similar construction but smaller. The two welded steel ribs, 3 m × 14 m, which form the arch are inclined inwards, and joined at the crown by massive steel plating

and circular distance pieces. At the ends they are 33 m between bearings. They are 2 m apart at the crown, and the height of the arch from the centre to rail level is 543 m. The two ribs straddle the deck, and rest on separate bearings on piers 6 and 7, as can be seen from fig 62. In fact the deck is only connected to the arch by the steel tie bars by which it is suspended, and it rests on its own separate bearings on the piers. In this respect it differs from the British Frodsham bridges where at the two ends, the arch and deck are integral, and rest on a common bearing.

The construction of the deck for the sixth span differs from that of the approach spans, and is shown in fig 64. It is 2 m deep with three longitudinal web plates one under each rail of the single track railway, and the third under the outside edge of the roadway. These web plates are cross connected together by main and intermediate cross girders. The main cross girders are spaced 11 m apart, and are deeper than the intermediate ones, and are extended

Plate 7.23 *Lorraine Viaduct, Berne, Switzerland*

beyond the deck width to provide anchorages for the inclined steel ties which support the deck from the arch. There are five intermediate cross girders between each pair of main ones. The whole deck is welded and bolted together using HSFG bolts. The tie bars which support the deck besides being inclined inwardly, are also inclined in the longitudinal direction, so that they form a double-Vee pattern, and each attachment position at the ends of the main cross girders has two ties secured to it. The ties are of different size, those on the railway side are 194 mm diameter, and those on the roadway side are 81 mm diameter. This is clear from fig 64, and also from the photograph of the bridge. The erection of the navigational span was carried out after the completion of the

approach spans on either side of it. Two steel jury masts suitably guyed were erected on the ends of spans 5 and 7 to support a construction cable way to enable the separate sections of the deck to be placed in position, and in addition temporary steel trestling was erected in midstream to support the crown of the arch. The arch was erected first followed by the deck sections and the attachment of the ties which were then tightened up to the correct degree.

The whole bridge is a notable example of modern steel and concrete construction.

There are several other examples of the 'tied arch' in Germany, notably, in the case of the Storstrombruke, where there are three arches at the centre of long approach viaducts. This bridge is located between Seeland and Falster.

Plate 7.24 *Grandfey Viaduct, Switzerland*

Plate 7.25 *Fehmarnsund Bridge, Germany/Denmark*

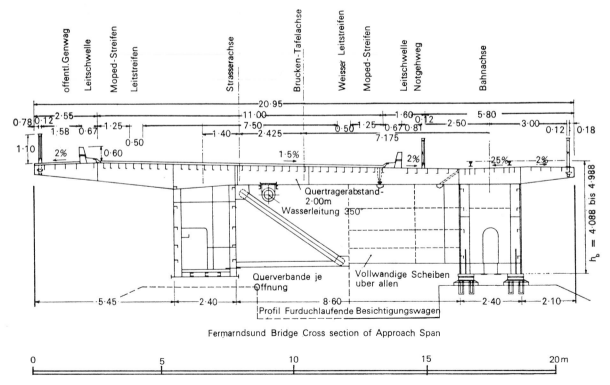

Fig. 62 *Fehmarnsund Bridge : detail of approach span*

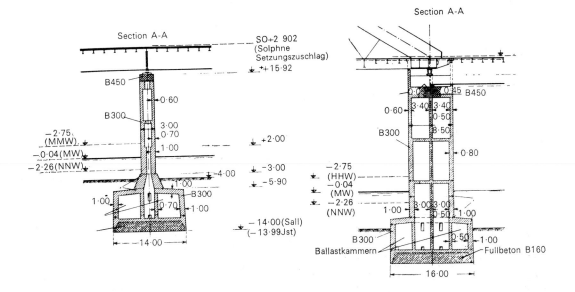

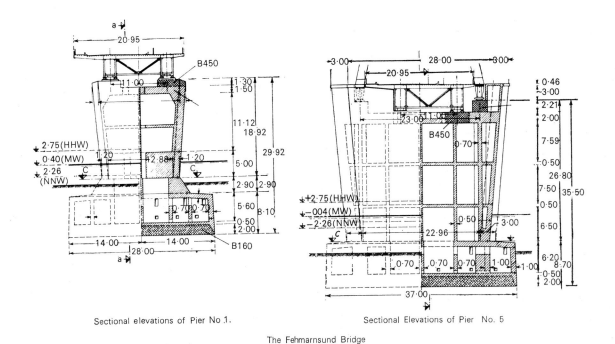

Sectional elevations of Pier No. 1.

Sectional Elevations of Pier No. 5

The Fehmarnsund Bridge

Fig. 63 *Fehmarnsund Bridge: detail of piers*

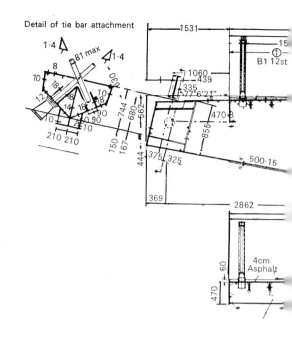

Detail of tie bar attachment

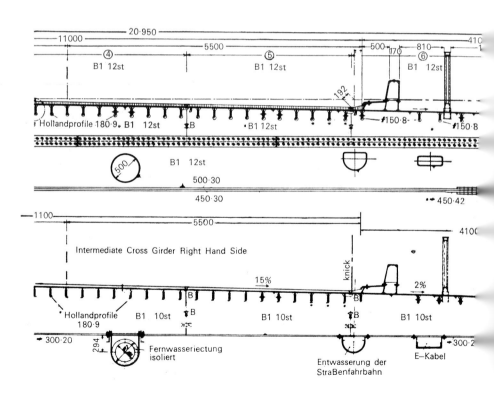

The Fehmarnsund Bridge Details

Fig. 64 *Fehmarnsund Bridge : detail of main span*

Main Cross Girder Left hand side

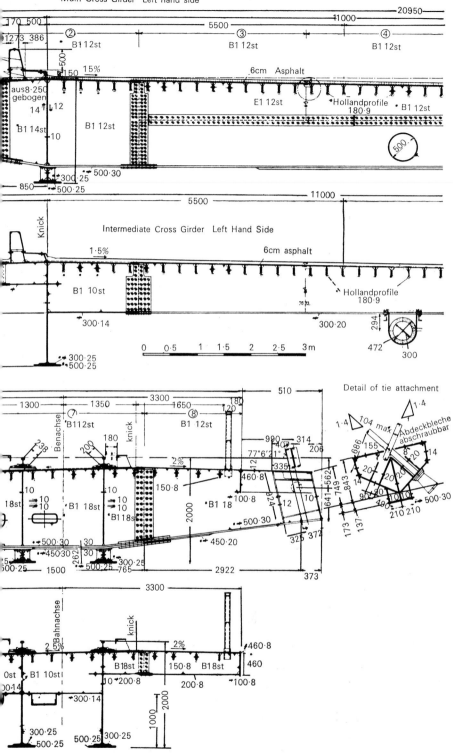

Intermediate Cross Girder Left Hand Side

Detail of tie attachment

Index

Abbey Wood Footbridge, S Region, 101
Abbots Langley Subway, LM Region, 22
Abeles, F W, 7
Adam Viaduct, LM Region, 57
Adamsdown Footbridge, Cardiff, W Region, 97
Alphon Brook Bridge, W Region, 22, 25
Anglian Products Ltd, 78
Arakawa Bridge, Japan, 130, 131, 134
Arlberg bridges, 134
Arrol, Sir William & Co, 43
Atherstone Bridge No 60A, LM Region, 12
Atkins, W S, Partners, 43, 81
Attock Bridge, Pakistan, 49, 124, 126, 127
Austrian Federal Railways, 134

Baes, Prof L, 12
Barnwell, F R L, CCE, W Region, 43
Beatty, W F, CCE, LM Region, 35, 69, 74, 92, 121
Belfast, 100, 101
Bennett, Hubert, Architect, 101
Berridge, P S A, xix
Besses o' th' Barn Bridge, LM Region, 19, 28, 63, 69, 72, 73
Bietschal Bridge, Switzerland, 146
Birmingham, New Street Station, 94
Bishops Stortford Footbridge, 101
Bletchley Fly-over Bridge, 80
Boot, Henry & Co, 37, 40
Boulton & Paul Ltd, 11
Britannia Bridge, xix
British Railways
    Eastern Region, 1, 7, 11, 19, 20, 37, 40
    London Midland Region, 13, 11, 12, 18, 19, 22, 24, 25, 35, 36
    North Eastern Region, 78
    Scottish Region, 7
    Southern Region, 1, 22, 24, 30, 32, 40, 78
    Western Region, 1, 22, 24, 25
    Research Department, Derby, 73

British Railways Board, 2
Brunel, Isambard Kingdom, xix
Buck Lane Bridge, E Region, 7
Burma, 127, 130
Butland, A N, CCE, LM Region, 59
Butterley Engineering Co Ltd, 37, 40, 49

Cantrell, A H, CCE, S Region, 35, 42
Caronte Swing Bridge, France, 141, 143
Cementation Ltd, 59
Central Africa, 122
China, 122, 123
Chislehurst Pipe Crossing, S Region, 115
Churchdown Bridge, W Region, 1, 43, 45, 46
Cleveland Bridge & Engineering Co Ltd, 122
Clifton Bridge (Penrith), LM Region, 25, 27, 28, 65, 67, 69
Concrete Development Association, 17
Concrete Services Ltd, 98
Congleton Footbridge, LM Region, 98
Conspan Ltd, 101
Costain, Richard, Ltd, 35
Costain Construction Ltd, 59
Crown Agents, 122
Croydon, Pipe Crossing, S Region, 115

Dannatt, H M, xvii
Denmark, 143
Deptford Creek Bridge, S Region, 40, 41, 42
Derby, Bridge No 1E, LM Region, 19, 24, 90, 91, 92
Dorman Long & Co Ltd, 34, 115, 122
Dow Mac Products Ltd, 99

Edgehill Cutting, Liverpool, LM Region, 94
Ellesmere Port Footbridge, LM Region, 100
Esk Viaduct, LM Region, 58
Evans, R E, CCE, S Region, 79
Everall, W T, 124

Exeter Footbridge, S Region, 99
Exeter pipe crossing, S Region, 115

Fairclough, Leonard Ltd., 67, 69, 90
Fairfield Street Bridge, Manchester, 18, 28, 59, 63, 72
Fehmarnsund Bridge, Germany, 147, 149, 150
Fledborough Viaduct, E Region, 49
Fletcher, Thomas, Ltd, Mansfield, 49
Footbridge No 89, LM Region, 100
Forth Bridge, xix
Freeman Fox & Partners, xx, 30, 32
French Railways, SNCF, 13, 138, 139, 140, 141, 143
Freysennet, 27
Frodsham Bridges, LM Region, 19, 24, 74, 76, 147

German 'Half Through' type Bridge, 7
Glacier Bearings, 17
Glasgow Suburban Electrification, 7
Goode & Partners, 122
Grandfey Viaduct, Switzerland, 146
Grosvenor Bridge, xix, 1, 30, 31, 32, 33, 34
Gstaad Viaduct, Switzerland, 146

Hadsphaltic Ltd, 49
Hampstead Road Bridge, LM Region, 94
Haresfield Bridge, W Region, 1, 43, 46, 47, 79, 91
Holland Hannen & Cubitts Ltd, 101
Howard Humphries & Sons, 101
Huntonbridge Subway, LM Region, 116
Husband, Messrs, Consultants, 76
Hyde Lane Bridge, W Region, 1, 43, 48, 49
Hyde Road Bridge, Manchester, LM Region, 18, 59, 62, 63, 64

Imperial College of Technology, 35, 64
India, 123, 124
Institution of Civil Engineers, 7, 60

Japan, Permanent Way Society, 134
Japanese National Railways, 130, 131, 134

Johnstan Bros Ltd, 48
Juggal River Bridge, India, 123, 124

Kander Viaduct, Switzerland, 146
Kirk Maclure & Morton Ltd, 101
Kleinenbelt Bridge, Denmark, 143

Landwasser Viaduct, Switzerland, 146
Leeds Canal Bridge, NE Region, 78
Lepski, A., 11
Lisbon, River Tagus Bridge, 137, 138
Liverpool St–Colchester, Clacton, Electrification, 7
Lorraine Viaduct, Berne, Switzerland, 146
Luogelkin Viaduct, Switzerland, 146

MacMurdo, A W, CCE, NE Region, 90
Manchester, Oxford Road, LM Region, 63
Manchester Piccadilly Station, 63
Manchester–Sheffield–Wath Electrification, 7, 12
Manchester South Junction and Altringham Railway, 18, 59, 63
Marazion Footbridge, W Region, 101
Marples Ridgeway Ltd, 35
Martin, D R & Sons Ltd, 101
Middlesbrough Pipe Crossing, 115
Ministry of Transport, 14, 20
Monk, A & Co Ltd, 92
Mossband Bridge, LM Region, 14, 15, 17
Mouselow Bridge, LM Region, 12
Mowlem Construction (Nairobi) Ltd, 122
Musgrave Hospital Footbridge, Belfast, 100, 101

Netherlands Railways, 19
New Pudsey Footbridge, E Region, 98
Newcastle on Tyne–Pilgrim St Bridge, 81, 82, 83, 84

Otztaler Bridge, Austria, 135, 136, 137
Oxford Road, Manchester, 63

Pakistan, 124, 125
Pakwash Bridge, River Nile, Uganda, 122
Partial Prestressing, 7
Penman, G., 93
Penrith Bridges, LM Region, 3, 19, 27, 65
Pinhoe Footbridge, S Region, 99
'Pipe-jacking', 21, 43, 115
Port of London Authority, 30
Portugal, 137, 138
'Preflex' Beams, 11
Prestbury Footbridge, LM Region, 100
Prestressed concrete, 7

Railway Gazette, 12
Rees, Wm J Ltd, 21, 115, 116
Rendel Palmer & Tritton, 7, 11, 49, 99, 124, 130, 146
Retford Dive-under Bridge, E Region, 37, 38, 39
Richmond Subway, S Region, 22, 116
Rolling bridges into position, 25
Roofing over cuttings, 94
Rowley Regis Bridge, LM Region, 26
Rugby Fly-over, LM Region, 80
Runcorn Footbridge, LM Region, 100
Runcorn Pipe Bridge, 92
Rungis Market Bridges, SNCF, France, 138, 139, 140

Sadler, R E, 7, 12
Schanatobel Bridge, Austria, 134, 135
Scott, Lawrence, Ltd, 42
Scratchwood Bridge, 11
Scunthorpe Footbridge, E Region, 98
'Seerthrust' system of 'pipe-jacking', 115
Sittang Bridge, Burma, 127, 130
Sliding bridges into position, 25, 26, 27, 28, 29
Skirsgill Bridge (Penrith), LM Region, 65, 67
Soil Mechanics Ltd, 91
Stephenson, Robert, xix
Stevenage Bridge No 85A, E Region, 19, 20, 28, 84, 85, 88, 89, 90

Stevenage Development Corporation, 84
Stewarts & Lloyds Ltd, 98
Stockport Road Bridge, Manchester, 18, 59, 60, 61, 62, 64
Surveyor, The, 115
Swiss Federal Railways, 143, 146

Tagus Bridge, Lisbon, Portugal, 137, 138
Tay Bridge, xix
Tees Dock Approach Road, 11
Teesside Bridge & Engineering Co Ltd, 34
Temple Mills Footbridge, E Region, 98
Tenby Footbridge, W Region, 101
Terris, A K, CCE, E Region, 37, 40, 49, 99
Thomson, J C, 116
Tinsley Canal Bridge, E Region, 1, 37, 39, 40
Tring Footbridge, LM Region, 100
Trowell Bridge, LM Region, 3, 35, 36
Tube Headings Ltd, 21, 22, 48, 80, 91, 116, 121
Tubewrights Ltd, 38
Turriff Construction Co Ltd, 48
Turton, F, 18, 60

Uganda, 122
United States Steel Corporation, 138

Wander Viaduct, Switzerland, 146
Wandsworth Bridge No 3A, S Region, 78, 79, 91
Watford Bridge No 66, LM Region, 12
Watford Vehicle Subway, LM Region, 116, 117
Welded steel construction, 1, 2
Wellerman Bros Ltd, 78
Westcliffe Footbridge, E Region, 101

Yangtze River Bridge, China, 122, 123
York Footbridge, NE Region, 100
Yorkshire Hennebique Ltd, 78